Lecture Notes in Computer Science 7836

Commenced Publication in 1973
Founding and Former Series Editors:
Gerhard Goos, Juris Hartmanis, and Jan van Leeuwen

Marcin Detyniecki Ana García-Serrano
Andreas Nürnberger   Sebastian Stober (Eds.)

Adaptive Multimedia Retrieval

Large-Scale Multimedia Retrieval and Evaluation

9th International Workshop, AMR 2011
Barcelona, Spain, July 18-19, 2011
Revised Selected Papers

 Springer

Volume Editors

Marcin Detyniecki
Université Pierre et Marie Curie
Laboratoire d'Informatique de Paris 6 (LIP6)
104, Avenue du Président Kennedy, 75016 Paris, France
E-mail: marcin.detyniecki@lip6.fr

Ana García-Serrano
Universidad Nacional de Educación a Distancia (UNED)
Departamento de Lenguajes y Sistemas Informáticos (LSI)
Ciudad Universitaria
28040 Madrid, Spain
E-mail: agarcia@lsi.uned.es

Andreas Nürnberger
Sebastian Stober
Otto-von-Guericke University
Faculty of Computer Science
Universitätsplatz 2, 39106 Magdeburg, Germany
E-mail: {andreas.nuernberger, sebastian.stober}@ovgu.de

ISSN 0302-9743   e-ISSN 1611-3349
ISBN 978-3-642-37424-1   e-ISBN 978-3-642-37425-8
DOI 10.1007/978-3-642-37425-8
Springer Heidelberg Dordrecht London New York

Library of Congress Control Number: 2013934555

CR Subject Classification (1998): H.3, H.5.1, H.5.5, H.2.8, H.5.2-3, H.2, H.4, I.2.6

LNCS Sublibrary: SL 3 – Information Systems and Application, incl. Internet/Web
and HCI

Typesetting: Camera-ready by author, data conversion by Scientific Publishing Services, Chennai, India

Printed on acid-free paper

Springer is part of Springer Science+Business Media (www.springer.com)

Preface

This book comprises a selection of revised contributions that were initially submitted to the 9th International Workshop on Adaptive Multimedia Retrieval (AMR 2011). This time the workshop was organized by the CLIC research group at the Universidad de Barcelona in Barcelona, Spain, during July 18-19, 2011.

The general goal of the AMR workshops is to intensify the exchange of ideas between the different research communities involved in this topic. To promote this exchange, some of the AMR events were collocated with artificial intelligence-related conferences, because different AI techniques contribute to the Adaptive Multimedia Retrieval Challenge. In 2011 the workshop was also organized as co-located event with the Joint Conference on Artificial Intelligence (IJCAI 2011).

Systems for searching and organizing multimedia information have matured during the last few years. However, retrieving specific media objects is still, a challenging task, especially if the query can only be vaguely defined or refers to different types of media. The main reasons for the problems in multimedia search are still, on the one hand, the users' difficulty in specifying their interests in the form of a well-defined query due to insufficient support from the interface, and, on the other hand, the problem of extracting relevant (semantic) features from the multimedia objects itself. Besides, ideally user-specific interests should be considered when ranking or automatically organizing result sets. To improve today's retrieval tools and thus the overall satisfaction of a user, it is necessary to develop advanced techniques able to support the user in the interactive retrieval process.

The book includes contributions ranging from theoretical work to practical implementations and its evaluation. This time, most of the papers deal with audio or music media, and different tasks are addressed: methodological evaluation of speech-retrieval systems, singing voice or music band member detection, fingerprinting-based or multimodal music retrieval. Regarding image retrieval, still a challenging topic, two papers are included that present two content-based techniques to improve the effectiveness of retrieval systems and a third one in which a new approach is taken in order to browse image or video collections. Especially here, the chosen similarity plays an important role in the underlying adaptation approaches as proven by the two experiments presented based on the MagnaTagATune benchmark dataset. Finally, in the special contribution of the invited speaker Stéphane Marchand-Maillet, approaches to tackle the scalability issues in multimedia retrieval by using distributed storage and computing architectures are discussed. The authors propose an approach that is able to handle interactive feedback and analyze its performance gain in terms of interactivity (search time) using a part of the Image-Net collection.

We would like to thank all the members of the Program Committee for supporting us in the reviewing process, the workshop participants for their willingness to revise and extend their papers for this book, the sponsors for their financial help, and Alfred Hofmann from Springer for his support in the publishing process.

September 2012   Marcin Detyniecki
                                                        Ana García-Serrano
                                                       Andreas Nürnberger
                                                          Sebastian Stober

Organization

Program Chairs

Marcin Detyniecki   CNRS, Laboratoire d'Informatique de Paris 6, France

Ana García-Serrano   Universidad Nacional de Educación a Distancia, Madrid, Spain

Andreas Nürnberger   Otto von Guericke University, Magdeburg, Germany

Technical Chairs

Sebastian Stober   Otto von Guericke University, Magdeburg, Germany

Monserrat Nofre   Universitat de Barcelona, Spain

Local Organization

M. Antonia Martí   Universitat de Barcelona, Spain

Mariona Taulé   Universitat de Barcelona, Spain

Program Commitee

Jenny Benois-Pineau	University of Bordeaux, LABRI, France
Stefano Berretti	Università di Firenze, Italy
Susanne Boll	University of Oldenburg, Germany
Jesús Chamorro	Universidad de Granada, Spain
Juan Cigarrán	Universidad Nacional de Educación a Distancia, Spain
Bogdan Gabrys	Bournemouth University, UK
Fabien Gouyon	INESC Porto, Portugal
Xian-Sheng Hua	Microsoft Research, Beijing, China
Alejandro Jaimes	Telefónica R&D, Spain
Philippe Joly	Université Paul Sabatier, Toulouse, France
Gareth Jones	Dublin City University, Ireland
Joemon Jose	University of Glasgow, UK
Peter Knees	Johannes Kepler University Linz, Austria
Stefanos Kollias	National Technical University of Athens, Greece
Stéphane Marchand-Maillet	University of Geneva, Switzerland
Trevor Martin	University of Bristol, UK

José María Martínez
  Sánchez   Universidad Autónoma de Madrid, Spain
Bernard Merialdo   Institut Eurécom, Sophia Antipolis, France
Gabriella Pasi   Università degli Studi di Milano Bicocca, Italy
Valery Petrushin   Accenture Technology Labs, Chicago, USA
Daniel Racoceanu   IPAL UMI CNRS 2955 Singapore
Stefan Rüger   The Open University, Milton Keynes, UK
Simone Santini   Universidad Autonoma de Madrid, Spain
Raimondo Schettini   University of Milano Bicocca, Italy
Ingo Schmitt   University of Cottbus, Germany
Nicu Sebe   University of Amsterdam, The Netherlands
Alan F. Smeaton   Dublin City University, Ireland
Arjen De Vries   CWI, Amsterdam, The Netherlands

Supporting Institutions

Universidad de Barcelona, Spain
Ministerio de Ciencia e Innovación, Spain
Otto-von-Guericke-University, Magdeburg, Germany
Laboratoire d'Informatique de Paris 6 (LIP6), France
Centre national de la recherche scientifique (CNRS), France
Universidad Nacional de Educación a Distancia, Madrid, Spain

Table of Contents

Invited Contribution

Evaluation and User Studies

Audio and Music

Image Retrieval

Similarity and Music

Learning-Based Interactive Retrieval in Large-Scale Multimedia Collections

Hisham Mohamed[1], Marc von Wyl[1],
Eric Bruno[2,*], and Stéphane Marchand-Maillet[1]

[1] Viper Group - Department of Computer Science - University of Geneva - CH
http://viper.unige.ch
[2] Data Mining and Knowledge Discovery - Corporate R&D Division - Firmenich SA - CH

Abstract. Indexing web-scale multimedia is only possible by distributing storage and computing efforts. Existing large-scale content-based indexing services mostly do not offer interactive relevance feedback. Here, we detail the construction of our Cross-Modal Search Engine (CMSE) implementing a query-by-example search strategy with relevance feedback and distributed over a cluster of 20 Dual core machines using MPI. We present the performance gain in terms of interactivity (search time) using a part of the Image-Net collection containing more than one million images as base example.

1 Introduction

Web-scale digital assets comprise millions or billions of documents. Professional search engines, to some extend, already cope with these scales. However, there is a lack of a proper common understanding of behaviours of content-based search multimedia engines at large scales. Academic content-based search engines coping with large-scale multimedia collections typically manage collections of millions of documents. The MUFIN search engine [2] proposes the indexing via M-trees of the CoPhIR image collection of 100 millions images. The BigImBaz system [11] indexes 10 million images. In both cases, the query consists of one positive example and search is mapped onto a neighborhood search of that example. This is the case also on most commercial services where similarity search is mapped onto a pre-indexed neighborhood search. The impressive feature here is the speed at which such queries are processed.

Here, we propose the design of a large-scale multimedia search engine accepting true relevance feedback from the user during the interactive search session. In section 2, we first briefly present our learning-based Cross-Modal retrieval approach [5] incorporating positive and negative relevance feedback over multiple examples. Then, in section 3, we investigate the distribution of the indexing and search operations over a computer cluster to preserve acceptable performance in the interactive setup. We finally propose an evaluation of our prototype in 4.

2 Learning-based Retrieval in Multimedia Collections

A *multimedia document* is composed of multimodal content (for instance visual, audio and textual content) and *multimedia information retrieval* consists of determining the

---

* This work was performed while the author was in the Viper group – University of Geneva.

M. Detyniecki et al. (Eds.): AMR 2011, LNCS 7836, pp. 1–17, 2013.

relevance of every document relatively to a given query. This relevance reflects the match of the multimodal content to the query.

Described briefly, our search algorithm presented next uses this user's supervision to create ranking for every chosen representative feature and every positive example. The non-linear combination of these base ranks is then done in a boosting-like fashion, thus privileging useful features and discarding redundant ones.

2.1   Preference-Based Representation

In the following, we consider a collection $\mathcal{X}$ containing l multimedia documents x_i, represented over m modalities (features). The terms item, element or object are also used to refer to x_i. The *query by example* search paradigm consists in collecting user's judgements indicating, for some objects, whether they are relevant or irrelevant to the user request. This set, denoted $\mathcal{Q}$, is called the *query* and is composed of positive and negative subsets, respectively

$$\mathcal{P} = \{x_i^+\}_{i=1}^p \text{ and } \mathcal{N} = \{x_i^-\}_{i=1}^n \text{ and } \mathcal{Q} = \mathcal{P} \cup \mathcal{N}$$

The query $\mathcal{Q}$ is then used to train a machine that will produce a decision function ranking documents according to their relevance to the query.

This paradigm might be embedded into the *Relevance Feedback* (RF) strategy, where these two steps (user judgement and ranking estimation) are iterated until the search converges to a satisfactory result. In [5], we proposed a *Query-based Dissimilarity Space* (QDS), derived from the dissimilarity spaces introduced by Pekalska *et al* [13].

The QDS presents two decisive advantages relatively to feature spaces: 1) It provides a unified representation of multimodal information channels, and 2) is particularly adapted to the class asymmetry typically exhibited by the positive and negative classes. This asymmetry corresponds to a $(1 + K)$ class setup where the one class, presumably well-clustered in the feature space, encompasses the sought documents (positive class), while an unknown number K of classes, partially represented by negative examples, is supposed to model all irrelevant documents. Classical learning approaches, by applying a symmetric treatment to all classes are not really efficient for such a setup. Learning the negative classes, while being feasible using traditional non-linear learning machines, becomes challenging when only few samples are available. Nevertheless, we show in [5] how a built-in property of our strategy is able to transform the asymmetric classification setup such that it becomes linearly separable. For more technical details about the internals of the search engine, the reader is referred to [5,3].

However, the issue of how to properly scale dissimilarity spaces so that modalities become easily comparable still remains. This problem might be left out to the fusion and ranking algorithms [5], but a more elegant solution would be to end up with a fully homogeneous multimodal representation.

In [4], we have proposed to simplify the QDS representation by replacing the dissimilarity components $d^k(x, x_i^+)$ with the ranking position $\pi^k(x, x_i^+) \in \mathbb{N}$ of an object

x with respect to the prototype x_i^+ according to the dissimilarity measure d^k relative to the feature k and the collection $\mathcal{X}$,

$$\pi^k(x, x_i^+) = \sum_{x_j \in \mathcal{X}} [\![d^k(x_j, x_i^+) \leq d^k(x, x_i^+)]\!]. \tag{1}$$

The notation $[\![\kappa]\!]$ is defined to be 1 if predicate κ holds and 0 otherwise. Considering the p positive prototypes and the m dissimilarity measures, the multimodal representation of an object x may be represented as a *unique* $(p*m)$-dimensional vector of *preferences*

$$\begin{aligned}
\boldsymbol{\pi}(x) = [\pi^1(x, x_1^+), \pi_1^2(x, x_1^+) \ldots, \pi^m(x, x_1^+), \\
\pi^1(x, x_2^+), \pi_1^2(x, x_2^+) \ldots, \pi^m(x, x_2^+), \\
\vdots \\
\pi^1(x, x_p^+), \pi_1^2(x, x_p^+), \ldots \pi^m(x, x_p^+)]^T.
\end{aligned}$$

For the sake of readability, the notation $\pi^k(\cdot, x_i^+)$ is simplified to $\pi_j(\cdot)$, $j = k + m * (i-1)$, $j \in [1, p*m]$, with i iterating over all objects $x_i^+ \in \mathcal{P}$ and k over the m modalities (features).

It is worth noting here that we obtain this modality-independent representation at the cost of losing most information about the initial feature distributions; only ordering information is actually preserved. Our objective now is to define a learning machine effectively able to learn from preferences as efficiently as learning directly in feature spaces or in dissimilarity spaces.

2.2 The Ranking Problem

The ranking problem may be formulated as follows: For each item $x \in \mathcal{X}$, it exists ranking features $\pi_1, \ldots, \pi_{pm}$, where each π_j defines a linear ordering of the instances $x \in \mathcal{X}$. In our formulation, $\pi_j \in \mathbb{N}$ and $\pi_j(x_1) < \pi_j(x_0)$ means that x_1 is preferred to x_0.

Additionally to the ranking features, there exists a *feedback function* $\Phi : \mathcal{X} \times \mathcal{X}$ which provides the learner with the desired form of the final ranking. Formally, $\Phi(x_1, x_0) > 0$ means that x_1 should be ranked above x_0 while $\Phi(x_1, x_0) < 0$ means the opposite. $\Phi(x_1, x_0) = 0$ means no preferences between x_0 and x_1 and the magnitude of $|\Phi(x_1, x_0)|$ indicates how important it is to rank x_1 above or below x_0. This function is said to be *bipartite* if there exists two disjoint set $\mathcal{X}_1$ and $\mathcal{X}_0$ such that Φ ranks all instances x_1 of $\mathcal{X}_1$ above instances x_0 of $\mathcal{X}_0$. These subsets are respectively the positive and negative subsets $\mathcal{P}$ and $\mathcal{N}$ we defined in section 2.

Learning such a feedback function implies estimating a ranking $H : \mathcal{X} \to \mathbb{R}$ through the optimization of a ranking loss function penalizing every miss-ordered pair of items. We consider the loss proposed in [9]

$$\sum_{\substack{x^- \in \mathcal{N} \\ x^+ \in \mathcal{P}}} \Phi(x^+, x^-) \left[H(x^+) - H(x^-) \right]. \tag{2}$$

The function $H(x)$ is a ranking of items x stating that x^+ is ranked higher than x^- whenever $H(x^+) > H(x^-)$. Interestingly, in case of bipartite feedback, the problem becomes separable and the ranking loss simplifies to [9]

$$\sum_{x \in \mathcal{Q}} w(x)s(x)H(x), \tag{3}$$

where the user feedback is carried by both

$$s(x) = \begin{cases} +1 & \text{if } x \in \mathcal{P} \\ -1 & \text{if } x \in \mathcal{N} \end{cases}, \tag{4}$$

and $w(x)$ a weight giving the importance of the rank of the item x.

Following the boosting principle, the final ranking H results from a weighted sum of weak rankings $h_t : \mathcal{X} \to \mathbb{R}$

$$H(x) = \sum_{t=1}^{T} \alpha_t h_t(x), \tag{5}$$

which is estimated through an Adaboost-like algorithm, namely RankBoost [9].

In essence, our algorithm thus estimates a global ranking function based on local rankers, that are themselves derived from comparing the intrinsic distance (in feature space k) between a positive example x_i^+ and any item x in the collection to access the values of $d^k(x, x_i^+)$. Since positive examples are supposed to be selected from the collection $\mathcal{X}$, one preprocessing could consist of calculating all pair-wise distances $d^k(x_i, x_j)$ and use feature-based distance matrices as index to the learners. This would result in a $O(m * l * l)$-sized storage, which is prohibitive and would not scale. Instead, we perform a reduction of the dimension of these distance-based representations using FastMap [8] and store all l items using f projections, leading to a $O(m * l * f)$-sized storage (for a typical small value of $f \ll l$) and an efficient distance computation. In practice, these projections are stored into an optimized SQLite table, which guarantees further fast access.

The special case of text is handled classically using the TF-IDF weighting scheme (with cosine similarity), in conjunction with Inverted Files.

3   Distributed Multimedia Information Retrieval

Despite the dimension reduction performed, a study of our search runs showed that most of the CPU effort is dedicated to the computation of distance values between items, i.e the comparison between each positive example and every image in the collection over each feature, in the first part of our RankBoost strategy. A multithreaded memory-shared approach may be sufficient to speed up the processing of a small collection but, for millions of elements, we face problems of memory storage and concurrent access to the mass storage (an SQLite database file in our case). The indexing of the ImageCLEF

Wikipedia collection comprising about 240'000 annotated images was possible on a 8-core machine with 16Gb of RAM, while preserving somewhat acceptable search times (about 10 seconds).

However, to move up scale and to cope with the problems of memory and mass storage access, we owe to consider more elaborated distributed computing architectures. This is clearly visible from the search times obtained with a sequential version (1 core only) over a large collection of images, as presented in section 4. We therefore first briefly review possible parallel architectures that may be considered (section 3) and then detail our informed choice for a cluster-based design of our architecture concerning both generic features (section 3.2) and text (section 3.3).

3.1 Parallel Computing Memory Architecture

Parallel computing is the execution of a base task on multiple processors or multiple computers, in order to obtain faster results by splitting up the problem into smaller subproblems.

There are many applications that require fast processing such as data mining (e.g. business, finance, multimedia), natural event (e.g. climate, earthquake, organ) simulation or engineering (e.g. car, circuit) design, simulation and test. Parallel computing models can be divided into essentially three categories: 1) Shared Memory Model, 2) Distributed Memory Model, 3) Hybrid Memory Model, which we discuss below.

Shared Memory Model. The shared memory model proposes a system that is composed of a number of independent cores (CPUs), where these cores share the same memory (Figure 1). Changes at a memory location made by one of the processors are visible to all other processors.

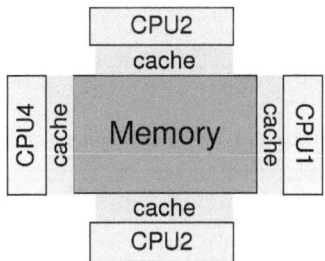

Fig. 1. Shared memory Model: 4 cores are located in the same machine, where every core manages a memory cache, and all cores access the same global memory storage

The best programming model for this architecture is threading model. If a machine hold N cores, it can initiate N parallel non-overlapping threads, each of which will resolve part of the problem. The most common libraries for multithreaded programming are OpenMP and PThreads.

The main advantage of this model is its ability to ensure fast sharing of data between processors. All cores see the same memory storage, so they do not need to communicate to share data between them.

In return, the corresponding main disadvantage is that this architecture is not easily scalable, as we cannot add more cores to the system easily.

Distributed Memory Model. This model envisages two or more machines. Each machine holds a processor, some local memory, and a network card (Figure 2). The machines are connected through a (supposedly fast) network communication. The processor within each machine has only access to its corresponding memory storage and cannot access the memory of the other processors, so changes made in a local memory have no effect on the memory of other processors. When a processor requires to access data managed by another processor, it is the programmer's task to define how and when data will be exchanged and communicated. For this, the most common distributed libraries are the MPI API (Message passing Interface) and the PVM library (Parallel Virtual machine).

The main advantages of this model are scalability and cost effectiveness. Thanks to the network versatility, to increase the computational power and the memory storage, one may easily add more nodes to the model. Also, it is a cost-effective solution since one can build a high performance computer using commodity, off-the-shelf processors and networking. This comes at the cost of complex programming and debugging operations, since independent entities need to be synchronised, either in terms of operations or in terms on the data they process. Further, the cost and delays incurred by network communication between the nodes affect the performance. Hence, adding more nodes to the computing network does not always guarantee an increase in the computing performance.

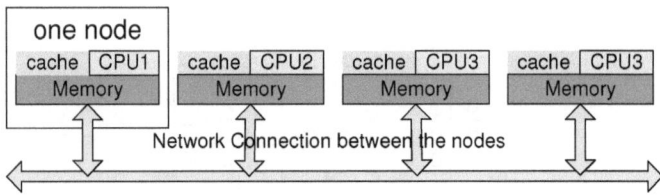

Fig. 2. Distributed Memory Model: 4 nodes, each one manages its own local memory and exchanging data is done through the network

Hybrid Memory Model. This model is a combination of both previous models. It recognises the generalisation of multi-core computers. It is essentially a Distributed Network of multi-core machines. Every multi-core computer is a node and nodes are connected through a communication network (Figure 3). Locally, the shared memory model is applied and this, changes made in local memory only affect core on the same node. Exchange of data between nodes is to be managed explicitly by the programming model.

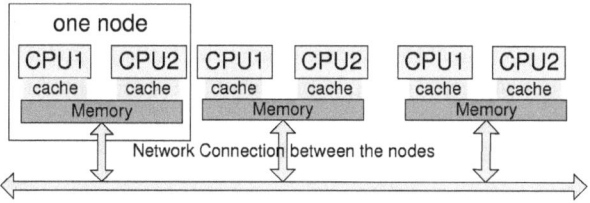

Fig. 3. Hybrid memory Model: 3 nodes, each node has its own memory and 3 cores, exchanging data between nodes is done through network connection

Hence, relevant programming languages are the same as for the above models and hold the same pros and cons.

3.2 Parallel Feature-Based Search

In section 2, we have detailed our learning-based search strategy. It is typically based on the combination of independent weak rankers combined into a global ranking function (equation (5)). This model readily enables a suitable distribution policy for the indexing of visual features. The systematic computation of distance values between every positive example document and the rest of the collection with respect to a given feature may be split over a distributed network of computing nodes, following the Distributed Memory Model.

Every node manages its own database containing a part of the collection. Every node is used to make distance-based ranking of its part of the collection with respect to a feature and a given example. It then sends back this ranking to the master node that is in charge of running the central part of the RankBoost strategy (aggregation and final sorting). Using this strategy, we divide the effort by the number of available CPUs and also share the memory and mass storage access, at the cost of communicating resulting ranks between the nodes and the master process. We have studied a number of communication policies, essentially varying the grain of the messaging to optimise the communication efficiency. Using common index knowledge between the nodes and the master, we could make the communication cost acceptably cheap.

Indexing. The possible strategies for the distribution of the data (and thus the strategy for sharing the labour over nodes) are guided by the dimensions of the data. Within our model, we have m feature over l items.

Since feature representations may be processed independently, distributing the data with respect to the features seems to be a strategy of choice. Here, each node will process the inter-distance (and possibly ranking) between *all* items within its designated feature space. However, the number of features is typically smaller than the available number of nodes. Hence, to attain an acceptable average node occupation, several nodes will be affected to one single modality. This forces the splitting of the item list over a subset of nodes.

For example, say we have l items, m features, and n nodes. Then each node will be affected $O(l * m/n)$ item-feature representations and be dedicated to the computation of their distance from any given query, with respect to a given modality. Figure 4 shows the main idea of index distribution.

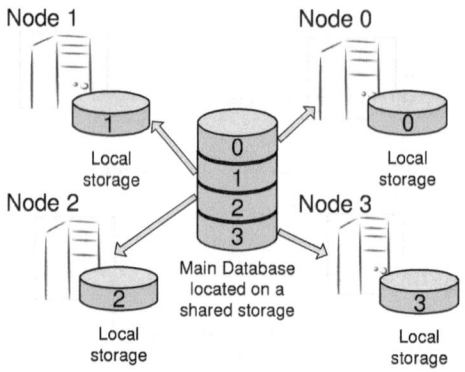

Fig. 4. Each node will be responsible for a part from the original collection of items

Parallel Search. Our model comprises a master node and a number of slaves associated to a part of the collection, as described above. When the master node receives the query from the user, it starts extracting the features from the query, so as to obtain a presentation that is comparable to that of the rest of the items. After that, it broadcasts the query features to the slave nodes. Every node in the cluster, including the master node itself, will then start computing query-item distances according to its responsibilities (i.e. within its own database segment). When the node is done, it sends the ranked results back to the master node. After receiving all the results from all the slaves, the master node sorts the results and return them to the server for display to the user (Figure 5).

Relevance Feedback on a Parallel System. After the results are displayed, the user selects the most relevant items to the current query, then sends again these items to the master node. According to our learning model, the design of the global ranking function, ranking all items with respect to all modalities, is a combination of weak ranks over all modalities with respect to each positive example. Hence, when the number of positive items fed back to the system increases, the computing time increases. However, since the relevance feedback is provided by the user, the number of positive example is typically small (typically around 10). Hence, this dependence is sustainable in terms of scalability.

3.3 Parallel Text Retrieval

In this study, we see text as a specific medium whose intrinsic properties may be exploited. Text indexing further bears a large literature whose findings are exploited here.

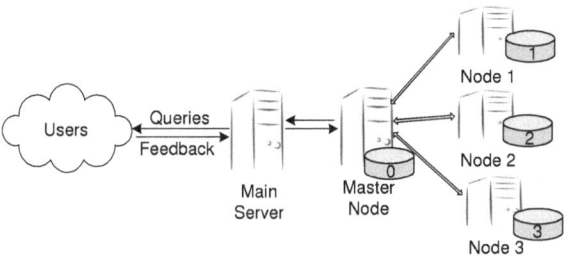

Fig. 5. Parallel Search: The master node distributes the query to the slaves and the replies with the search results

Indexing. Inverted files are the most used indexing technique for text collections [15]. An inverted file is composed of essentially two elements: the dictionary and the inverted list. The dictionary is a lexicographically sorted list containing a selection of unique terms appearing in the documents of the collection. Every term is then associated with an inverted list (posting list) containing the references to the documents where this term appears. Further, for each document, a weight is given that reflects the importance of the term into the document, in the context of the complete collection. The *TF-IDF* weighting scheme [15] is the most often used criterion in information retrieval and text-mining. This weight is a statistical measure used to evaluate how important a word is to a document in a corpus.

Due to the large volume of text data generally considered, which may include billions of documents, the process of building the inverted list may be a complex problem facing scalability issues. In order to preserve efficient processing and acceptable processing times, one should use parallel and distributed algorithms, which divide themselves into essentially two categories: local indexing and global indexing [1].

Local Indexing. The most direct way to build the inverted files for the documents is to divide the complete collection into sub-collections and distribute these over the machines. Each machine will thus be responsible for constructing an inverted file for its sub-collection. This strategy however affects the computation of the inverse document frequency (IDF) factor [1], since every node is not able to see the rest of the collection, which is distributed over the other nodes.

Global Indexing. The alternative is to build a global inverted file for the whole collection and then distribute this global inverted file over the nodes. Every node then receives a part of the global inverted file. The most common distribution strategy is based on the lexicographical order. Every node manages the inverted lists for terms starting within a given set of characters. For example, Node 1 is associated with letters "A", "B", "C"; Node 2 with letters "D", "E", "F"; and so on.

According to the existing literature, the global indexing strategy is the best choice to make [1]. The main challenge is then how to distribute the global inverted list. The baseline is to construct the complete global inverted list using one node only, then distribute the terms, term by term. This strategy is very slow and not commonly used due to the large volume of text data generally considered. The best way is thus to build the global inverted list and distribute it at the same time [12].

Distributed Global Text Indexing in Map-Reduce. Our distributed text indexing algorithm is based on single pass indexing and the Map-Reduce programming model. Single pass indexing [10] is the most recent technique in text indexing, where the posting list for every dictionary term is built in memory as the corpus is scanned. Since text corpora may be very large, the posting lists cannot be assumed to fit into the memory. Hence, whenever all the memory is consumed, the partial posting list is stored onto the disk and, once the processing of the whole collection is achieved, the stored segmented posting lists are concatenated.

The Map-Reduce programming model [6] is composed of two functions, namely "Map" and "Reduce". The *Map* function accepts input key-value pairs and emits key-value pairs: $M : (K_m \times V_m) \mapsto (K_r \times V_r)$. The *Reduce* function accepts a key and a list of values and generates a single return value; $R : (K_r \times V_r^N) \mapsto V$. After the mapper nodes emit the results, the Reduce function starts to group (e.g, sum) all the results pairs by their keys.

For text indexing, we have implemented Map-Reduce framework using MPI [14] (see Figure 6).

The general idea is to build the inverted files on the fly without the need to access the hard-disk in parallel on distributed machines. In our implementation, we have two types of nodes: the mappers and the reducers. The mappers are responsible to read documents,tokenize the documents and lexicographically sort the tokenized terms (dictionary terms) from every document. Mapper nodes emit key/value pairs. The key is, in this case, the term extracted from the file and the value is the document name and the corresponding TF value for the term:

$<k_m,$ $v_m>$ = <term,<document name, tf>>.

The key/value pairs are further sent to the reducing nodes. Each reducing node only receives the terms located into its lexicographic range. As the same words from all documents are saved into the same database, reducer nodes can calculate the correct value of the IDF and then assign a weight to every term according to the TF-IDF scheme.

Retrieval. At the end of the indexing operations, the database is distributed on number of nodes based on its lexicographic ordering. Upon receiving the query under the form of a document, the server transmits it to the master node. The master node then processes the query and extracts appropriate terms. After sorting the terms, the master node sends every term to the relevant nodes, ie the nodes associated with a lexical range known to contain the term in question.

The slave node then retrieves the documents most relevant to the term it was queried with and sends the results back to the master node. Since the data is distributed by terms, several query terms may retrieve more than one instance of the same document, each associated with respective weighting related to this term. The master node will therefore gather messages and compute the weighting value of the each individual document and finally sort the resulting relevant documents according to this final weighting.

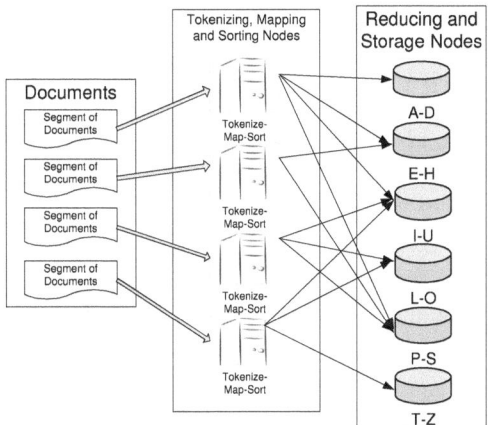

Fig. 6. Implementation 2: The nodes are divided into two categories: mapper nodes and reducing nodes

4 Experiments and Results

We have conducted large-scale experiments to test the validity of our distributed retrieval models. The results section is divided into two parts. We have indexed 1,091,860 randomly selected images from the 12-million image ImageNet corpus [7] to test our query by example scenario. Our goal is to later increase the number of indexed images and index the complete ImageNet corpus. Then, we have indexed 9,319,561 text (XML) excerpts related to 9,319,561 images from within the same ImageNet corpus.

4.1 Experimental Platform

The Cross Modal Search Engine (CMSE) is a C++ library including easy-to-use and extensible feature extraction, indexing and search capabilities. In its current version, it proposes visual feature extraction (color, shape, texture, face), dimensionality reduction of the extracted features using the FastMap algorithm [8] and text indexing with TF-IDF weighting embedded into an inverted file strategy. Feature extraction and dimensionality reduction are offline operations feeding an SQLite 3.0 database. The actual time-critical operation is the online search with relevance feedback. Similarity computation between the combination of positive and negative examples and every image in the collection is based on the RankBoost learning strategy (see section 2).

In order to handle large-scale problems, we have moved onto a cluster-based distributed strategy, corresponding to the above hybrid parallel processing model. For this purpose, we use the C++ message passing library MPICH2, an open implementation of the MPI API. We have thus modified the initial sequential and multithreaded version to install it onto a cluster of 20 DualCore computers (40 cores in total) holding each 8GB of memory and 512GB of local disk storage, led by a master 8-core computer holding 32GB of memory and a TeraByte storage capacity.

4.2 Query by Example

In this section, we will measure the response time of the system for the first query and for the user feedback. We are based on the multimodal indexing of 1,091,860 randomly selected images from the 12-million image ImageNet corpus [7]. Features include Grid Color Moments, Texture histogram, shape analysis and face detection. The distribution of the computational effort is as described in section 3.2.

Query Response Time. As described earlier, when a query is received by the master node, it distributes it onto the cluster nodes, each processing the query against its share of the global corpus.

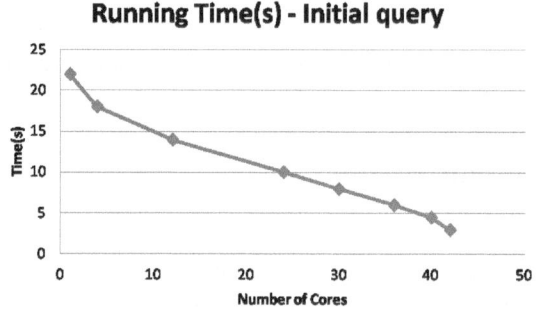

Fig. 7. Influence of the number of available cores for processing the first query step

Figure 7 shows the response times for processing the initial query (ie, without user feedback) based on an increasing number of cores. For each subset of cores used, we have indexed the complete database based on this number of available cores, as if the cluster comprised these cores only. At every stage, we submit the same query and make sure we receive the same results. Here, we submit an initial query formed of only 1 positive example image, corresponding to a click-through ("More like this") scenario.

We see that the processing time decreases as the number of available cores augments. We note that the gain is not linear and this is mostly due to the communication time that increases as the number of cores augments. Naturally, as we are based on a star-shaped connection model, as the number of cores augments, the load for the exchange of data between cores increase, leading to a raw parallel computing cost.

We finally note that using our complete cluster of nodes leads to acceptable query processing times (< 5 seconds for more than 1 million documents).

4.3 User Feedback Response

Whenever the user provides feedback, the response time is affected in essentially two ways. First, for every positive example image provided, the engine must compute its distance to the rest of the collection, according to all features (section 2). Figure 8 compares response times in function of the number of available cores, when providing 5 or 10 positive and negative examples images in a feedback step (4th step in this case).

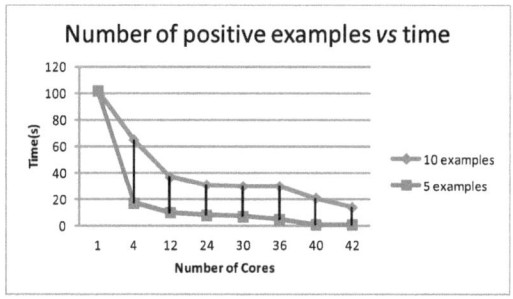

Fig. 8. Influence of the number of examples in the feedback steps

Second, according to our iterative learning model, the more coherent will the example images provided be, the faster the convergence of the learning operations. Hence, it is expected that processing times for later feedback steps (where the exploration is localised into a small neighborhood of the feature space) will be faster than for early feedback steps (where examples may be less coherent). Figure 9 shows measured response times for 5 feedback iterations when providing 5 (left) and 10 (right) positive and negative example images at every step. The running time decreases when the number of nodes increases and also when the number of iterations increases.

Note that in our experiments, we used automatically generated feedback, whose coherence may be lower than human-provided feedback. Hence, in practice, these measured running times may be considered as unfavourable cases.

4.4 Further Performance Factors

Memory usage, Speedup and Efficiency are other main parameters to measure the performance of any parallel system.

Memory Usage. We have first measured the size of the memory allocated in function of the number of iterations on different number of nodes. Figure 10, shows the memory usage for different number of iterations and different number of positive examples.

Clearly, when the number of iterations and the number of positive examples increase, more memory is consumed. At the same time, the memory usage by each core decreases when the number of cores increase.

Speedup and Efficiency. Speedup and Efficiency respectively measure the gain distributing operations. Speedup (S) computes the raw gain against a sequential system and Efficiency (E) estimates the average occupation of each core of the system.

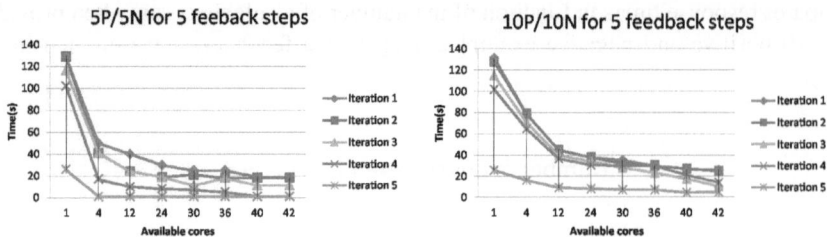

Fig. 9. Right: The response time for 5 iterations for 5 positive images feedback in each iteration. Left: The response time for 5 iterations for 10 positive images feedback in each iteration.

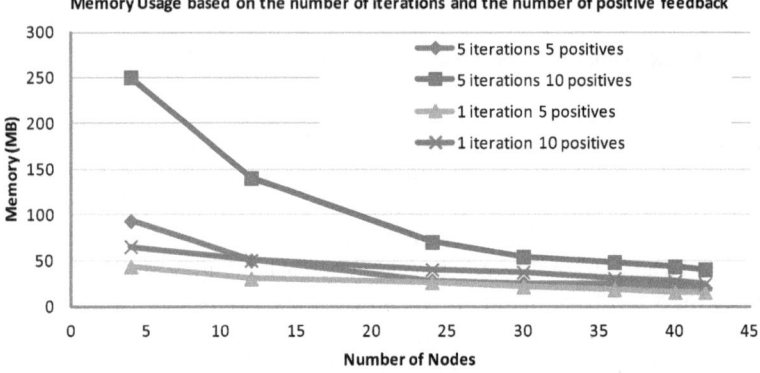

Fig. 10. Memory consumption

$$S = T_s/T_p \text{ and } E = S/P$$

where P is the number of cores in the cluster, T_s is the sequential execution time and T_p is the parallel execution time. Ideally, Speedup should be linear in function of the number of available cores and Efficiency should remain close to 1. Figure 11 shows the Speedup and Efficiency measured as the number of nodes increases.

While the expected trend of a linear Speedup is obtained, the Efficiency is far from optimal with a value of 20%. This shows that the machines do not work effectively together and that the load is not optimally balanced during the complete process. This is clearly a place where there is room for improvement in our distribution strategy. However, one also should consider that multiple queries may be processed simultaneously and it is not clear whether reaching and 100%-Efficiency (leading to processing queries sequentially) is a goal to achieve.

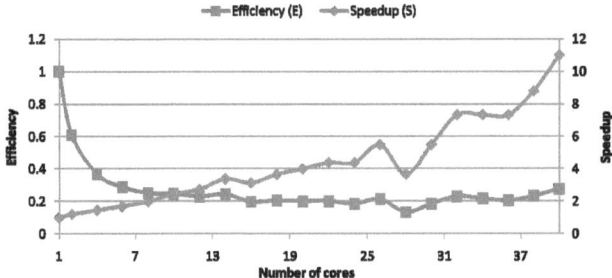

Fig. 11. Speedup and Efficiency

4.5 Text Query

We now turn onto text processing and retrieval. We evaluate our model proposed in section 3.3. We first evaluate text indexing based on our strategy when varying the size of the available cluster. Then, we validate the efficiency of our model in performing efficient retrieval over a large corpus of text. In our context, text consists of 9,319,561 short text excerpts (XML files) attached to images in the ImageNet corpus, leading to the selection of 18,675 unique terms in our dictionary.

Indexing Performance. We measure the performance when varying the parameters of our model, namely, the number of mapper nodes, the number of reducing nodes, and the number of available cores in the cluster.

The number of mapper nodes directly affects the processing time since it induces the partition of the collection over the cluster. In other words, when fewer mapper nodes are declared, each of these nodes will receive a larger set of documents to handle. Figure 12 plots the total indexing time based on the number of mappers for 2, 4, 13 and 26

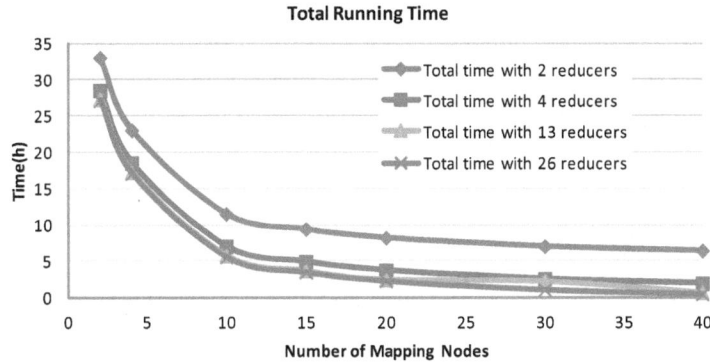

Fig. 12. Total indexing time with respect to number of mapper and reducer nodes

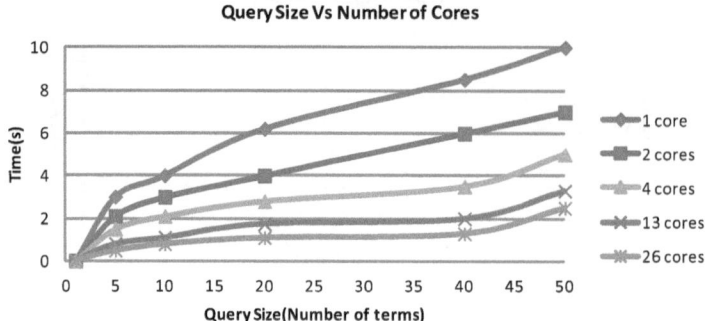

Fig. 13. The running time with respect to different queries size

reducer nodes. Clearly, when the number of mapper and the reducer nodes increase the running time systematically decreases, with a non-linear decay.

Sequential time (one node for all the processing) was experimented on two different types of nodes. One with 32GB of memory and the other one with 8GB of memory. In the first case, the indexing time took 17.5 hours, which is very long comparing to our distributed performance. In the second case, the memory could not accommodate all of the data and the system needed to write to the swap. The running time at this case was in excess of 8 days. In both cases, we demonstrated a true gain of a distributed model over a sequential processing.

Searching Time. Here, the critical factor is the length of the query (ie number of terms in the query), inducing more or less processing over the nodes of the cluster. Figure 13 shows the obtained running times for different query sizes, on cluster of various sizes. We chose different queries with different sizes from the dataset, then applied the same query on different distribution of the database, respectively over 2, 4, 16 and 26 nodes.

Again here, there is a true gain in distributing the data over a large cluster of nodes.

5   Conclusion

Managing large volumes of multimedia data imposes to share the storage and processing over multiple processors, both for indexing and query processing tasks. We propose a distributed learning-based multimedia information retrieval model that is based on the systematic comparison of every positive example with the rest of the collection according to available features. We have detailed the distribution of indexing and query processing operation over an hybrid parallel computing architecture.

We have tested our proposal over our cluster of 40 cores and showed that this architecture indeed leads to significant gains in terms of processing times. Not only this is beneficial for the complete process but it also allows to match user's requirement in acceptable waiting times (generally few seconds when using 40 cores) and thus install a true relevance feedback mechanism that we see as a way to reach acceptable retrieval performance.

We proposed here initial performance measures and gain based on processing times and memory occupation. We will further conduct experiments to understand better factors such as parallel query processing, load balance and communication times.

Our framework currently distributes the work load based on exact retrieval. Scalability is this restricted by the linear dependence between the number of documents in a collection and the processing power of each core. We are currently looking at how approximate search may remove this dependence while preserving retrieval quality.

Acknowledgements. This work is jointly supported by the Swiss National Science Foundation (SNSF) via the Swiss National Center of Competence in Research (NCCR) on Interactive Multimodal Information Management (IM2) and the European COST Action on Multilingual and Multifaceted Interactive Information Access (MUMIA) via the Swiss State Secretariat for Education and Research (SER).

References

1. Badue, C., Baeza-yates, R., Ribeiro-neto, B., Ziviani, N.: Distributed query processing using partitioned inverted files. In: Proc. of the 9th String Processing and Information Retrieval Symposium (SPIRE), pp. 10–20. IEEE CS Press (2001)
2. Batko, M., Falchi, F., Lucchese, C., Novak, D., Perego, R., Rabitti, F., Sedmidubsky, J., Zezula, P.: Building a web-scale image similarity search system. Multimedia Tools and Applications 47(3), 599–629 (2010)
3. Bruno, E., Kludas, J., Marchand-Maillet, S.: Combining multimodal preferences for multimedia information retrieval. In: Proceedings of the International Workshop on Multimedia Information Retrieval (2007)
4. Bruno, E., Marchand-Maillet, S.: Multimodal preference aggregation for multimedia information retrieval. Journal of Multimedia 4(5), 321–329 (2009)
5. Bruno, E., Moënne-Loccoz, N., Marchand-Maillet, S.: Design of multimodal dissimilarity spaces for retrieval of multimedia documents. IEEE Transactions on Pattern Analysis and Machine Intelligence 30(9), 1520–1533 (2008)
6. Dean, J., Ghemawat, S.: Mapreduce: simplified data processing on large clusters. Commun. ACM 51, 107–113 (2008)
7. Deng, J., Dong, W., Socher, R., Li, L.-J., Li, K., Fei-Fei, L.: ImageNet: A large-scale hierarchical image database. In: IEEE Computer Vision and Pattern Recognition (CVPR) (2009)
8. Faloutsos, C., Lin, K.-I.: Fastmap: a fast algorithm for indexing, data-mining and visualization of traditional and multimedia datasets. SIGMOD Rec. 24(2), 163–174 (1995)
9. Freund, Y., Iyer, R., Schapire, R.E., Singer, Y.: An efficient boosting algorithm for combining preferences. Journal of Machine Learning Research 4, 933–969 (2003)
10. Heinz, S., Zobel, J.: Efficient single-pass index construction for text databases. J. Am. Soc. Inf. Sci. Technol. 54, 713–729 (2003)
11. Schmid, C., Jégou, H., Douze, M.: Improving bag-of-features for large scale image search. International Journal of Computer Vision 87(3) (2010)
12. McCreadie, R., Macdonald, C., Ounis, I.: MapReduce indexing strategies: Studying scalability and efficiency. Information Processing and Management (2011)
13. Pekalska, E., Paclík, P., Duin, R.: A generalized kernel approach to dissimilarity-based classification. Journal of Machine Learning Research 2, 175–211 (2001)
14. Squyres, J.M.: Definitions and fundamentals – the message passing interface (MPI). Cluster World Magazine, MPI Mechanic Column 1(1), 26–29 (2003)
15. Witten, I.H., Moffat, A., Bell, T.C.: Managing Gigabytes: Compressing and Indexing Documents and Images, 2nd edn. Morgan Kaufmann, San Francisco (1999)

A User Study of Visual Search Performance with Interactive 2D and 3D Storyboards

Klaus Schoeffmann, David Ahlström, and Laszlo Böszörmenyi

Klagenfurt University,
Universitaetsstr. 65-67, 9020 Klagenfurt, Austria
{klaus.schoeffmann,david.ahlstroem,laszlo.boeszoermenyi}@aau.at

Abstract. A storyboard is a grid-like arrangement of images, or key-frames of videos, that is commonly used to browse image or video collections or to present results of a query in an image or video retrieval tool. We investigate alternatives to the commonly used scroll-based 2D storyboard for the task of browsing a large set of images. Through a user study with 28 participants we evaluate three different kinds of storyboards in terms of visual search performance and user satisfaction. Our results show that a 3D cylindrical visualization of a storyboard is a promising alternative to the conventional scroll-based storyboard.

1 Introduction

A storyboard is a grid-like arrangement of images, typically with a scrolling or paging function, that allows a user to interactively inspect and browse a set of images. Storyboards have been intensively used for many years in the field of image and video retrieval as they enable users to browse through a collection of images or through the results of a query. While already introduced almost two decades ago [1], the traditional scroll-based 2D storyboard is still used by many state-of-the-art image and video retrieval tools. For example, most systems in the interactive Known Item Search (KIS) task in TRECVID 2010 [22] used scrolling- or paging-based 2D storyboards for interactive browsing.

However, these storyboards have limitations when browsing a large set of images (or key-frames of shots of videos). For example, at any time users can see only an excerpt of the whole set of images, which makes it difficult to visually compare images located in different parts of the list and to remember images that have been seen already. With many images in the list it is also hard for a user to estimate the total number of images (can be deduced from the size of the scrollbar). In addition, when scrolling through the whole list without finding the desired image, it is quite frustrating to go back in reverse order or to completely scroll up in order to restart from the beginning. More importantly, with common storyboards there is no convenient and intuitive way of inspecting the current excerpt of images at high level of detail while keeping remaining images in view, maybe with lower detail.

The interactive KIS task in TRECVID [22] simulates a use case where a user needs to find a specific video within a very large collection. In a first step

M. Detyniecki et al. (Eds.): AMR 2011, LNCS 7836, pp. 18–32, 2013.

the retrieval system performs some pre-processing of videos in the archive (in TRECVID according to a text query) and then presents the user with a result list of shots that is most likely to contain the desired content (out of 146,768 shots in about 9,000 videos). Each shot in that result list is typically represented by a single key-frame that is often used as link/button to start playback for the shot. In the second step the user performs an interactive visual search in this rather long result list (300 results are quite usual) in order to find the single video of interest. As pointed out earlier, this is an inconvenient process for the user and the question arises if the conventional two-dimensional storyboard is the best interface to support such an interactive visual search process.

In this work we investigate three different storyboard interfaces for the purpose of interactive visual search in a long list of images or key-frames of shots, such as required by the interactive KIS scenario in TRECVID. The three interface compared are (i) a complete grid of small images without scrolling, (ii) a scrolling-based storyboard, and (iii) a 3D cylindrical storyboard. For all interfaces we use roughly the same screen estate but different presentation and navigation methods. We report results from a user study with 28 users, which compares both the performance in task completion time and subjective workload ratings (experienced convenience, mental workload, physical workload etc., as defined by the NASA Task-Load-Index [11]). We show that a novel 3D storyboard has great potential to allow for more convenient and more flexible interactive visual search than conventional storyboards.

2   Related Work

Several investigations for the presentation and visualization of images and key-frames of shots have been performed in the past. Komlodi and Marchionini [12] compared storyboards with dynamic approaches (such as slide-shows) and report that storyboards were still the preferred method, even if additional time is required to interact with the user interface (scrollbars) and for visual search. Dynamic approaches often display the content with a fixed frame rate and do not allow the user to adjust the speed of progress. Graph-based approaches for visualization of multimedia content can also be found in the literature. For example, Delest et al. [8] used spatio-temporal color signatures for graph clustering in order to visualize the shot structure of a video, according to shot-based similarity. Also Rodden et al. [17] investigated ways of organizing images of large collections in a storyboard-like arrangement by visual similarity. They compared random organization with organization by visual similarity for image selection tasks (e.g., selection of photos from three different places/locations). Rodden et al. found that users were slower with the similarity-based organization. As potential explanation, they pointed out that an arrangement based on visual similarity may increase the probability that the searched image is completely overseen because surrounding images make it less eye-catching. Moreover, their user study did also show that in the case when both arrangements are available most users (21 out of 30) tend to use the similarity-based arrangement first, but like to have both arrangements available. Through a user simulation, Nguyen and Worring [14] showed that for very large image collections visualization schemes based on visual similarity can outperform storyboard-like visualizations for specific interactive search tasks. Visualization by similarity for interactive search was also investigated by Plant and Schaefer [15], who showed with a subjective test in a further study [18] that a 3D globe visualization of visually similar images is preferred by users over a conventional storyboard-like arrangement. Advanced similarity-based visualization approaches for images were also investigated by Quadrianto et al. [16]. In the literature only a few approaches can be found that take advantage of 3D graphics. In an early work, Manske [13] proposed to use a *Cone-Tree* representation of key frame hierarchies. He also investigated a *side-view* perspective to enable viewing of several transformed key-frames at a glance. A user is able to open and close sub-trees in order to inspect specific segments in more detail and to switch to a front-view perspective based on a selected key-frame. Moreover, a feature for an automatic *walk through* the whole video-tree in its temporal sequence is supported. Unfortunately, no user evaluation has been done, which would assess the performance of such a content representation. Divakaran et al.[9] proposed several techniques for improved navigation within a video through advanced content presentation with rather simple 3D transformations. For example, the *Squeeze* layout and the *Fisheye* layout have been presented for improved fast-forward and rewind with personal digital video recorders. With a user study Divakaran et al. showed that their approach can significantly outperform the VCR-like navigation set in accuracy. However, no significant difference was found in the task-solve time. The MediaMetro interface proposed by Chiu et al. [2] uses a 3D city metaphor for browsing video collections. Videos are visualized as skyscrapers on a grid layout with key frames shown on different sides of the building. Furthermore, each building/video has a single representative key frame, the "rooftop-image". The user can move around in the city with different navigation buttons, which is similar to flying with a helicopter. In order to support users in navigating through the city, a ground navigation on an animated path can be enabled by a mouse click at a point on the ground. A mouse click on any key frame/building will start a video player in an external window. The MediaMetro interface has been extended in a later work [3] to enable browsing of videos in a multi-display environment with a tabletop display and a vertically mounted flat panel. The user can interact with a gamepad that should facilitate navigation through the artificial city. de Rooij et al. proposed the *RotorBrowser* [7,6] that uses a 3D-like visualization of key-frames for browsing shots of videos in a video collection. The user starts browsing with an initial query that specifies textual and semantic components. The system computes relevant shots for the query and visualizes these shots through key-frames in the interface. The interface consists of several navigation paths, called *video threads*, that allow for exploration of shots that are similar to a focal shot S, which specifies the current position in the collection – at the beginning the most relevant shot for the query. Video threads are available for (i) shots with similar textual annotations, (ii) shots with similar

low-level visual features, (iii) shots with similar semantics, (iv) shots in chrono-
logical order of the video (timeline), and (v) shots in a ranked list of query
results. The video threads are shown as different directions in a Rotor-layout,
where the focal shot is always shown in the center. Navigation is possible through
the selection of any visible shot or by hotkeys. Recently, Christmann et al. [4]
suggested to use a cylindrical 3D visualization of large picture collections and
hypothesize that in terms of location memorization, viewing the cylinder from
the inside will be favorable over a perspective from the outside. They also spec-
ulate that an inner view should provide a more intuitive and richer experience
than when rotating the cylinder seen from an outside perspective. Furthermore,
they hypothesize that the outer view, which distorts larger items more than
smaller items, will result in better visual search efficiency than the inner view
where the smaller items are distorted more than the larger items. The results
show no difference in terms of item location learning and memorization, nor in
terms of search efficiency. But a marginal difference regarding perceived quality
of interaction experience, with a slight favor for the inner view. Although no direct comparison was made with a standard 2D interface, participants frequently
reported positive attitudes and possible advantages with the 3D visualization.
The *VisionGo* interface proposed by Zheng et al. [23] does also use the idea of a
partial vertical cylinder that is somehow similar to the interface proposed in this
work. However, to best of our knowledge the VisionGo interface does not use real
3D visualization with perspective distortion. No projects for general browsing
of images or key-frames of videos can be found in the literature that compare
2D interfaces with interfaces taking advantage of a 3D graphics environment
in order to visualize a storyboard-like layout that allows for more flexible and
convenient interaction and that provides high utilization of the screen estate (as
discussed in Section 3.3). Furthermore, it is not clear yet whether a complete list
of results in a simple grid-presentation of very small images will work better or
worse than a common scrolling-based storyboard. To our knowledge, the study
on Space Filling Thumbnails by Cockburn et al. [5] is the only previous study
that compares a dense interface with small-sized items to a scrolling interface.
Their results indicate favorable aspects for all-at-once approaches with miniatur-
ized items. However, the study used grayscale items (pages of text documents)
and the scrolling interface displayed only one item at a time (i.e., one page).
Thus, it is questionable whether and how the results can be generalized to a storyboard scenario with color images and scrolling interfaces displaying more than
one item at a time. In this paper we compare three alternatives through a user
study: (i) a complete grid of small images without scrolling, (ii) a scrolling-based
storyboard, and (iii) a 3D cylindrical storyboard.

3   Storyboard Interfaces

As an initial investigation of storyboard alternatives, the goal of this work is to
compare storyboard interfaces that are basically very similar. We are especially
interested in the following research questions:

1. Does a 3D storyboard outperform a 2D storyboard in terms of visual search time and/or subjective rating?
2. Do storyboard interfaces with very small images impede user performance?

The rationale behind the last question is the fact that 3D interfaces show content at different sizes (and different distortions) due to 3D projection. Therefore, for this initial investigation we decided to test three interfaces that differ in size of displayed images and in interaction possibilities. Only one interface uses 3D projection, whereas a perspective has been chosen that produces only minor distortion. All interfaces use approximately the same screen estate. This enables a fair comparison of different alternatives to be used as image/key-frame browsing approach in an image or video retrieval system. We focus on the interactive KIS task in TRECVID [22]. Therefore, our data set consists of 1,100 semi-randomly selected keyframes of shots from the IACC.1 collection of TRECVID 2010, from which 256 random images are used for every trial in the evaluation. The 1,100 selected shots represent a very diverse content set and contain no duplicates, nor highly similar images. Note that results of a real KIS query would contain highly similar keyframes of shots, which are ranked according to some semantic criteria like speech transcripts. This ranked list of images typically contains highly similar visual content too. However, we do not use real results of KIS queries as we want to avoid situations where the search task degenerates to a tedious comparison of unapparent image details. In all trials the target image is constantly shown on the left area in order to simulate the KIS scenario, where a user searches for a certain image in mind. For every interface the size of the target image was chosen twice the width and height of the average size of images.

3.1   Scroll (Common Scrollable Storyboard)

The *Scroll* interface, shown in Figure 1, is the common scroll-based storyboard that displays images at a convenient size and allows to scroll up or down. The interface shows 65 images at a glance. In our implementation, scrolling is possible by mouse-wheel interaction but no scrollbar is shown. The resolution of an image is 110×64 pixels.

3.2   Matrix (Complete Thumbnail List)

The *Matrix* interface, depicted in Figure 2, displays all 256 images in the result set in one screen. The resolution of an image is 55×32 pixels, which is half the width and height used in the Scroll interface. While the image size is very small, this interface has the advantage that no scrolling is required.

3.3   3D (Cylindrical 3D Storyboard)

In order to provide a 3D alternative to the common storyboard we investigated several different arrangements and interaction models that have been pre-evaluated through experimental tests. We found that a cylindrical alignment of

Fig. 1. Scroll Interface, 256 images in total

a common storyboard, which is actually a wrapped 2D storyboard, provides a very convenient interface [20] that provides the following advantages over the 2D storyboard (compare with Figure 3): (i) Images in focus are presented at large size while many of the remaining images are still visible, although at smaller size. (ii) When presenting several categories of content (e.g., semantic concepts), we can easily show one cylinder per category and enable the user to walk/zoom to a specific cylinder. (iii) The user can see a large amount of images in the collection and can quickly zoom/navigate to a different location in the cylinder. (iv) The cylinder can be automatically rotated, which gives a nice animated presentation of all images without any interaction. (v) The cylindrical arrangement works for a reasonably high and variable number of images. The more images in the collection, the larger the radius of the cylinder or the taller the cylinder (i.e. more 'rings'). Through usage of a 3D graphics environment the user can switch to different views (e.g., to a *front-view* as shown in Figure 4) or to quickly zoom/navigate to a certain area in the cylinder. (vi) In a typical content-based image/video retrieval tool the query results are arranged in row-major order according to rank. In such a scenario the 3D cylinder enables users to immediately continue search from the high-ranked images in case the end of the list was reached without success.

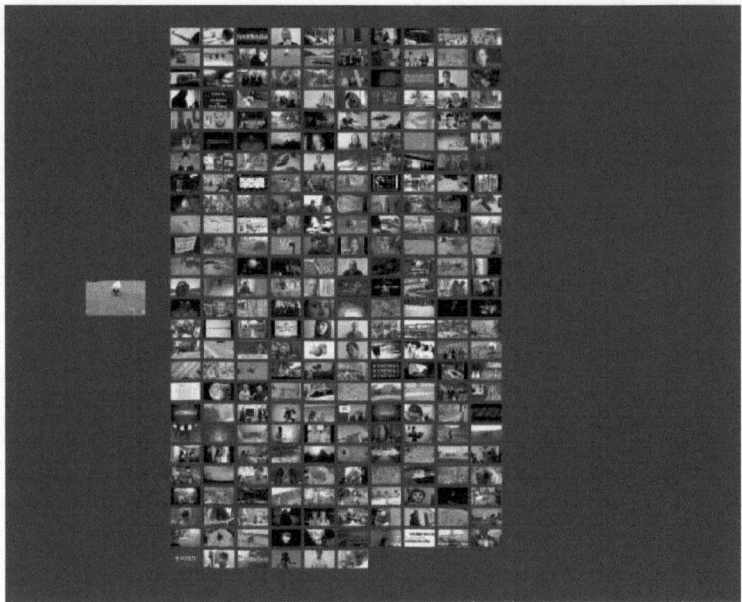

Fig. 2. Matrix Interface, 256 images

This novel storyboard is actually an elliptical cylinder that can be rotate in both directions by using the mouse-wheel. The height and width of the ellipse not only define how visible or occluded the images in the curvilinear storyboard are, but also have an influence on the rotation speed of images at specific areas in the curvilinear storyboard. In our implementation images farther away from the users' viewpoint move slower at rotation than images in close proximity. Moreover, we use a continuation of rotation with decreasing velocity, which means that the cylinder continues to rotate even if the user stops mouse-wheel interaction. However, this continuing rotation immediately slows down and halts after a short time. This feature is used to facilitate interaction with the rotating cylinder and to avoid extensive mouse-wheel usage. A user is, however, able to avoid this continuation by clicking the right mouse button, which instantly halts the cylinder.

However, for an initial evaluation of this interface we intended to provide such a perspective of the cylinder that produces minor distortion of images in order to keep the users' cognitive load as low as possible. Therefore, we decided to perform the first user test with the perspective shown in Figure 5 as this inner perspective minimizes distortion and might be more intuitive, as already mentioned by Christmann et al. [4] (see Section 2). Future user tests will be performed with more advanced perspectives and integrated experiences of this work. The resolution of an image in the cylinder as used in this first evaluation ranged from 74×50 pixels at the top of the screen over 110×64 pixels at the center of the screen to 196×136 pixels at the bottom of the screen. The real

Fig. 3. 3D Cylinder: *right-side view* with 300 images

resolution of the images (as extracted from the videos) is 430×240 pixels, which results in about 300KB in uncompressed RGB color space. The images are used as texture for '*3D screens*', each consisting of four vertices that are automatically scaled (and distorted) according the observers' position (the camera position). The interface shows about 80 partially occluded images at a glance, which is about 23% more than in the 2D scroll interface.

4   User Study

A user study was conducted to compare the efficiency of the three interfaces in a visual search task, according to interactive KIS in TRECVID.

4.1   Participants and Apparatus

Twenty-eight volunteers (fourteen female) aged 20 to 39 years (mean 27.39 years, SD 5.66) participated in the experiment. All participants used a computer and mouse on a daily basis with a self-estimated computer usage per week ranging from 9 to 80 hours (mean 40.35 hours, SD 19.27). All participants were right handed and were paid €10 in compensation for their participation. All three interfaces were tested on a Dell Precision M4400 Laptop (running Windows 7)

Fig. 4. Flexibility of 3D Cylinder: *front-view* with 300 images

with an external 24-inch display at a resolution of 1600×1200 pixels. A laser mouse (Dell LaserStream) was used as input device. The experiment software was coded in C# .NET 4.0 with the Microsoft XNA Framework 4.0.

4.2   Task and Procedure

The experimental task was to visually search for a cued image as fast as possible in a display of 256 randomly ranked images. This high number of images was chosen because in the TRECVID interactive KIS task it is not uncommon that a user has to search through a few hundreds of images. The target cue was displayed at a fixed position on the otherwise empty screen. A click on the cue started timing and the search display with its images was shown. A click on any image in the search display stopped timing, emptied the search display, loaded and showed a new target cue for the next trial. The experiment was administered in three blocks of trials, one block of 16 trials for each of the three interfaces (i.e., 48 trials per participant). The search display was divided into 16 equally large conceptual target groups of 16 images with grouping following a row-major order. Within each block, each of the 16 target groups was used once in random order and the exact target position within a target group was randomly chosen for each trial. This guaranteed an even distribution of target positions across the 256 possible positions in the search display. An error was logged when the wrong image was selected, and erroneous trials where put back at a random position in the queue of unfinished trials for the running block.

The images in the search display were randomly drawn from a pool with 1,100 images (the images were taken from the IACC.1 repository of TRECVID 2010) for each trial. After completion of a trial, all images but the target image were replaced into the pool, guaranteeing that a particular image was only used as target once within a block. A test session lasted 45-60 minutes and started with an approx. 10 minute introduction, explaining the task and demonstrating

Fig. 5. 3D Interface, 256 images in total

the three interfaces. After the introduction, participants had an open learning phase to allow them to familiarize themselves with the task and the interfaces. The trial block for the first interface was then presented and participants were asked to complete the trials as fast and accurately as possible and to take short breaks between trials as needed. After the completion of each interface participants answered a questionnaire about the perceived workload. Then the session progressed with a new learning phase and a block of trials with the next interface (presentation order for the interfaces was counterbalanced using a balanced Latin square). Finally, overall preference data and personal information were collected. In summary, the experiment consisted of 28 participants × 3 interfaces (Matrix, Scroll, 3D) × 16 trials = 1344 correct trials.

4.3   Measures, Analysis and Results

Trial time was calculated as the elapsed time from a click on the target cue to a click on any image in the display. Error rate was the percentage of trials within one block in which an incorrect image was clicked. Perceived workload measures used Likert scale ratings based on the NASA Task-Load-Index workload index [11] and included mental demand, physical demand, temporal demand, effort, performance and frustration. Trial time data was analyzed using a one-way repeated measures Analysis of Variance (ANOVA). Non-parametric Friedman tests were used on error rate and Likert scale data. Wilcoxon Signed-Rank tests with Bonferroni adjustment were used for post-hoc pair-wise comparisons of error rates and Likert scale data (the Bonferroni adjustment resulted in a reduced significance level set at p<.0167).

4.4   Error Rate and Trial Time

The overall error rate was 3.9% (55 trials from a total of 1399 trials). The error trials were evenly distributed across interfaces with 19, 17 and 19 error trials for Matrix, Scroll and 3D respectively. From the error free data, a total of 32 trials (2.4%) were identified as outliers (trial time > 3 SD of the interface mean time) and were removed from the temporal analysis (Matrix 12 trials, Scroll 9 trials, 3D 11 trials removed).

The overall mean trial time (errors and outliers removed) was 19.3s (SD 18.8). Mean trial time was 20.4s (SD 22.8) for the Matrix, 19.2s (SD 17.1) for the Scroll interface and 18.2s (SD 15.8) for the 3D interface. An RM-ANOVA showed, however, no significant difference in trial time between the three interfaces ($F_{2,54} = 1.212, p = .306$).

4.5   Speed-Accuracy Tradeoff

Cognitive tasks, such as our, are often subjected to a speed-accuracy trade-off with participants either focusing on accuracy or on speed (see Förster et al. [10] for an overview). Plotting the number of errors logged in the three interface for each participant against participants' mean trial times (error trials and outliers removed) does not reveal any general pattern indicating interface dependent speed-accuracy trade-off issues.

4.6   Workload Measures

Perceived workload measures are shown in Figure 6. We found that interface type had a significant impact on mental demand (χ^2(2,N=28)=10.84, p<.01), effort (χ^2(2,N=28)=10.98, p<.01) and frustration (χ^2(2,N=28)=12.73, p<.01) but not on physical demand (χ^2(2,N=28)=0.08, p=.961), temporal demand (χ^2(2,N=28)=5.01, p=.079) and performance (χ^2(2,N=28)=2.41, p=.299). Post-hoc pair-wise comparisons showed that the Matrix was more mentally demanding, effortful and frustrating than both the Scroll and 3D interfaces (all p's<.0167). There were no differences between the Scroll and the 3D interface regarding mental demand, effort and frustration.

Participants were also asked to indicate which of the three interfaces they liked the most and the least. The results are shown in Figure 7. More than half of the participants (15 from 28) ranked the 3D interface first (Rank 1) and 19 from 28 participants liked the Matrix interface the least (Rank 3). Overall, there was a significant difference in the preference rankings (χ^2(2,N=28)=13.50, p=.0012) and post-hoc analysis showed that there were no significant differences between the ratings of the Scroll and Matrix interfaces (Z=-2.236, p=.0253), or between the 3D and Scroll interfaces (Z=-1345, p=.1787), but a significant difference between ratings of the 3D and Matrix interfaces (Z=-3.311, p=.0009).

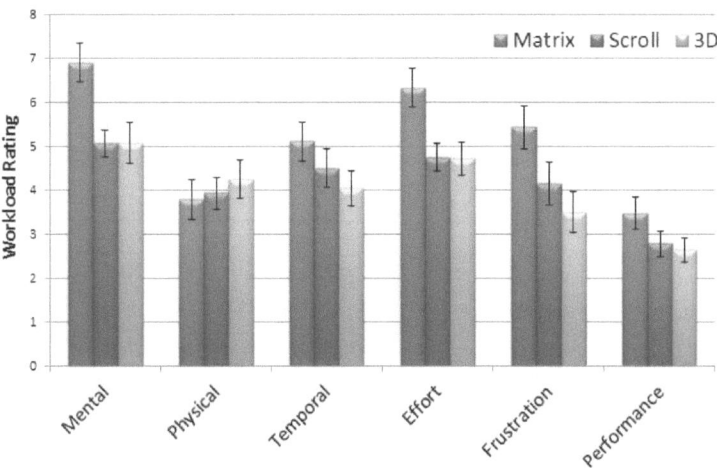

Fig. 6. Perceived workload ratings (lower ratings are better). Error bars represent 2 standard error (N=28).

5 Discussion

The results of this first user study show that there is no significant difference in terms of search time between the three tested interfaces, although the 3D interface achieved the best mean average value. However, from the subjective ratings of the perceived workload we can see that the Matrix interface performs significantly worse than the others in mental demand, effort, and frustration workload. Users obviously consider mental requirements and convenience as more important than achievable task solve time. In fact, the subjective ranking of interfaces (Figure 7) was only consistent with the measured ranking in terms of mean search time (including errors and outliers) for 12 out of 28 subjects (42.85%). In other words, it seems that users consider workload, such as mental demand or frustration for example (see Section 4), as more important than achievable search time. Four subjects rated the interface with which they were fastest as being the worst interface, which in all cases was the Matrix interface. More interestingly, 35.70% of users gave the 3D interface a better rank than measured in mean search time (only 7.14% of users gave a worse rank). The same situation was observed for the Scroll interface for only 17.80% of users (28.57% of users gave a worse rank).

It should be noted that this paper presents the results of our very first investigation of alternative interface to 2D storyboards. We used a very fair evaluation where each interface used roughly the same screen estate and the result lists did not contain similar images. The latter mentioned configuration was required to perform fair time measurements for search tasks and to avoid that a search task degenerates to a tedious comparison of unapparent image details. Therefore, our study should be considered as a general evaluation of interfaces for KIS tasks.

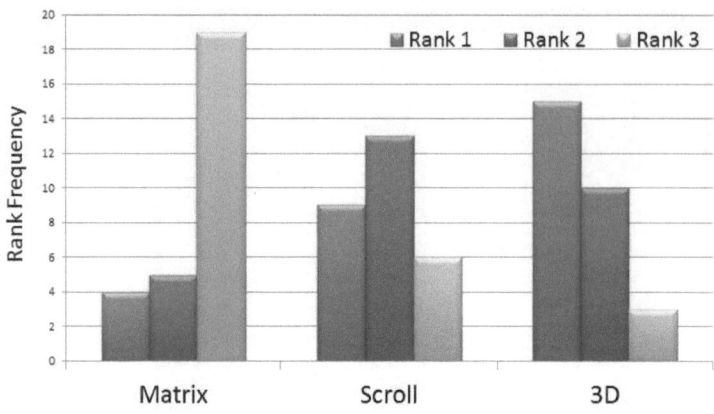

Fig. 7. Rank frequency

The results may differ for single tasks when similar results are in the search collection. It is however hard to evaluate those situations because the number of similar images in the results of the preprocessing step for a KIS task may heavily differ from one task to another.

All users were able to efficiently interact with the cylindrical 3D storyboard after a very short introduction although only half of them had ever used any kind of 3D software (mainly games) prior participating in the user study. This reflects the simple and intuitive interaction mechanisms of the 3D interface. It is interesting to note that the average subjective rating of the 3D interface was best for all but one workload measure, which was the physical demand. From personal interviews with participants we learned that the reason for the observed slightly higher physical demand of the 3D cylinder was the configuration of the scrolling speed of the cylinder, which was set rather low/fine (turning the mouse-wheel one notch moved the images in the cylinder approx. half a row).

It is also important to note that while the conventional 2D storyboard has specific limits, the 3D storyboard opens many avenues for improvement. For example, we can show more images at a glance (like in Figure 3) and provide a perspective zooming function (right mouse button). We could also improve the 3D visualization by using more than two triangles in order to show bended images instead of flat images. Our experiments have shown that such a visualization results in a more intuitive use of the cylinder and provides less occlusion [21]. We could provide shortcuts for switching the current viewpoint (front, left-side, right-side, center, etc.) that will enable fast inspection of several areas in the cylinder. Moreover, it seems that the 3D cylinder is particularly well suited for touch-based interaction and small display devices, such as tablet computers [19]. Also, instead of only one cylinder we could use several cylinders (e.g., with frontal view like in Figure 4) at a glance to present the image collection. For instance, each quarter of the whole collection could be represented by a cylinder. A scroll interaction in that scenario would rotate all four cylinders at once and therefore allow for parallel inspection of images in all quarters. 3D projection would allow to focus on a specific area/cylinder, while keeping in view the other images at lower detail.

6   Conclusions

Our results have shown that in terms of search time there is no statistically significant difference between all interfaces. However, we have also shown that there is a positive attitude for the 3D Cylinder (average subjective rating was best for all but one workload measure, see Fig. 6) and that the Matrix interface performs significantly worse in mental demand, effort, and frustration workload. From the evaluation results of the Matrix and the 2D Scroll interface we can further conclude that the much smaller size of images in the Matrix interface does not significantly affect visual search time. This encourages our intent to continue investigation of the 3D Cylinder interface with different and probably more advantageous perspectives. Finally, it should be noted that users in this first study were confronted with the 3D search interface for the very first time, while they were already very familiar with the Scroll interface. Therefore, we expect better results in search time with trained users and different perspectives, as we will investigate in future studies.

References

1. Arman, F., Depommier, R., Hsu, A., Chiu, M.: Content-based browsing of video sequences. In: Proceedings of the Second ACM International Conference on Multimedia, pp. 97–103. ACM Press (1994)
2. Chiu, P., Girgensohn, A., Lertsithichai, S., Polak, W., Shipman, F.: Mediametro: browsing multimedia document collections with a 3d city metaphor. In: Proceedings of the 13th Annual ACM International Conference on Multimedia, MULTIMEDIA 2005, pp. 213–214. ACM, New York (2005)
3. Chiu, P., Huang, J., Back, M., Diakopoulos, N., Doherty, J., Polak, W., Sun, X.: Mtable: browsing photos and videos on a tabletop system. In: Proceeding of the 16th ACM International Conference on Multimedia, MM 2008, pp. 1107–1108. ACM, New York (2008)
4. Christmann, O., Carbonell, N., Richir, S.: Visual search in dynamic 3d visualisations of unstructured picture collections. Interacting with Computers 22(5), 399–416 (2010); Modelling user experience
5. Cockburn, A., Gutwin, C., Alexander, J.: Faster document navigation with space-filling thumbnails. In: Proceedings of the SIGCHI Conference on Human Factors in Computing Systems, CHI 2006, pp. 1–10. ACM, New York (2006)
6. de Rooij, O., Snoek, C.G.M., Worring, M.: Mediamill: semantic video search using the rotorbrowser. In: Proceedings of the 6th ACM International Conference on Image and Video Retrieval, CIVR 2007, p. 649. ACM, New York (2007)
7. de Rooij, O., Snoek, C.G.M., Worring, M.: Query on demand video browsing. In: Proceedings of the 15th International Conference on Multimedia, MULTIMEDIA 2007, pp. 811–814. ACM, New York (2007)
8. Delest, M., Don, A., Benois-Pineau, J.: Dag-based visual interfaces for navigation in indexed video content. Multimedia Tools and Applications 31, 51–72 (2006), 10.1007/s11042-006-0032-4

9. Divakaran, A., Forlines, C., Lanning, T., Shipman, S., Wittenburg, K.: Augmenting fast-forward and rewind for personal digital video recorders. In: Proceedings of the IEEE International Conference on Consumer Electronics, ICCE, pp. 43–44. IEEE (2005)

10. Förster, J., Higgins, E., Bianco, A.: Speed/accuracy decisions in task performance: Built-in trade-off or separate strategic concerns? Organizational Behavior and Human Decision Processes 90(1), 148–164 (2003)

11. Hart, S., Staveland, L.: Development of NASA-TLX (Task Load Index): Results of Empirical and Theoretical Research. Elsevier B.V. (1988)

12. Komlodi, A., Marchionini, G.: Key frame preview techniques for video browsing. In: Proceedings of the 3rd ACM Conference on Digital Libraries, pp. 118–125. ACM Press (1998)

13. Manske, K.: Video browsing using 3D video content trees. In: Proceedings of the 1998 Workshop on New Paradigms in Information Visualization and Manipulation, pp. 20–24. ACM Press (1998)

14. Nguyen, G., Worring, M.: Interactive access to large image collections using similarity-based visualization. Journal of Visual Languages & Computing 19(2), 203–224 (2008)

15. Plant, W., Schaefer, G.: Visualising image databases. In: Proceedings of the IEEE International Workshop on Multimedia Signal Processing, MMSP 2009, pp. 1–6. IEEE (2009)

16. Quadrianto, N., Kersting, K., Tuytelaars, T., Buntine, W.: Beyond 2D-grids: a dependence maximization view on image browsing. In: Proceedings of the International Conference on Multimedia Information Retrieval, pp. 339–348. ACM Press (2010)

17. Rodden, K., Basalaj, W., Sinclair, D., Wood, K.: Does organisation by similarity assist image browsing? In: Proceedings of the SIGCHI Conference on Human Factors in Computing Systems, pp. 190–197. ACM Press (2001)

18. Schaefer, G.: A next generation browsing environment for large image repositories. Multimedia Tools and Applications 47(1), 105–120 (2010)

19. Schoeffmann, K., Ahlström, D., Beecks, C.: 3d Image browsing on mobile devices. In: Proceedings of the Seventh IEEE International Symposium on Multimedia, ISM 2011, Dana Point, California, USA (December 2011)

20. Schoeffmann, K., Boeszoermenyi, L.: Image and Video Browsing with a Cylindrical 3D Storyboard. In: Proceedings of the 2011 ACM International Conference on Multimedia Retrieval, ICMR 2011. ACM (2011)

21. Schoeffmann, K., del Fabro, M.: Hierarchical video browsing with a 3d carousel. In: Proceedings of the ACM International Conference on Multimedia, Scottsdale, AZ, USA (December 2011)

22. Smeaton, A.F., Over, P., Kraaij, W.: Evaluation campaigns and trecvid. In: MIR 2006: Proceedings of the 8th ACM International Workshop on Multimedia Information Retrieval, pp. 321–330. ACM Press (2006)

23. Zheng, Y., Neo, S., Chen, X., Chua, T.: Visiongo: towards true interactivity. In: Proceeding of the ACM International Conference on Image and Video Retrieval, CIVR 2009, pp. 51:1–51:1. ACM, New York (2009)

An Illustrated Methodology for Evaluating ASR Systems

María González[1], Julián Moreno[1], José Luis Martínez[2], and Paloma Martínez[1]

[1] Computer Science Department, Universidad Carlos III de Madrid,
Avda. Universidad 30, 28911, Leganés, Madrid, Spain
{mgonza1,jmschnei,pmf}@inf.uc3m.es
[2] DAEDALUS – Data, Decisions and Language S.A.
Avda. de la Albufera, 321
28031 Madrid, Spain
jmartinez@daedalus.es

Abstract. Automatic speech recognition technology can be integrated in an information retrieval process to allow searching on multimedia contents. But, in order to assure an adequate retrieval performance is necessary to state the quality of the recognition phase, especially in speaker-independent and domain-independent environments. This paper introduces a methodology to accomplish the evaluation of different speech recognition systems in several scenarios considering also the creation of new corpora of different types (broadcast news, interviews, etc.), especially in other languages apart from English that are not widely addressed in speech community.

Keywords: Automatic Speech Recognition (ASR), ASR Evaluation, Audio Transcription.

1 Introduction

One of the goals in current information retrieval research is going beyond text [1]. There is no doubt that users need to find different kinds of resources present in the web (audio, video, images) as well as using the same formats in their queries. So, multimedia formats are getting more attention, from video indexing to querying using images, audio, video or text. These formats can be applied to the information retrieval problem in different ways, from query by example of images or videos, to the conversion between formats, for example from video to image. Nevertheless, text representation is still the most representative one so many multimedia retrieval approaches are based on the use of metadata or on the transformation from any format to text. From this point of view, Automatic Speech Recognition (ASR) technology provides tools to transform human voice signals into text. Traditional text retrieval techniques can be then applied on the resulting text, providing good characterizations of multimedia objects. The transcription could be used to improve retrieval allowing extraction of relevant video and audio fragments concerning, for instance, keywords used in queries.

Nowadays, there are several ASR products available, from commercial ones such as Dragon Naturally Speaking (DNS) [5] or Microsoft Windows Speech Recognizer (WSR), to open source software packages like Sphynx [3] or HTK [4]. At this point,

M. Detyniecki et al. (Eds.): AMR 2011, LNCS 7836, pp. 33–42, 2013.

an important issue arises, which of these products best suites information retrieval system needs? There have been great efforts in ASR evaluation frameworks, particularly with some conferences devoted to ASR evaluation but they are, in general, designed from the point of view of final applications such as those promoted by TC-STAR[1] (Technology and Corpora for Speech to Speech Translation) focusing on Speech-to-Speech Translation or the Spoken Document Retrieval Task promoted by TREC[2] (Text Retrieval Evaluation Conference) in late 90´s or CL-SDR (Cross Language Speech Document Retrieval) from 2003 to 2007 launched by CLEF[3] (Cross Language European Forum). More recently, MediaEval Benchmark[4] 2011, an initiative for multimedia evaluation, includes two speech related tasks: Spoken Web Search Task and Rich Speech Retrieval Task. All of them are devoted to do IR from transcripts of spoken documents. As far as the authors of this paper know, there are not available ASR evaluation platforms allowing a comparison of several ASR products using different types of corpora in different scenarios.

Therefore, the availability of speech corpora is a central issue due to the difficulties and the cost of collecting and manually annotating a corpus with transcriptions [13]. The main corpora containing transcriptions in these tracks are: (1) American-English news recordings broadcast by ABC, CNN, Public Radio International, and Voice of America collected by the Linguistic Data Consortium (LDC) for ASR training and (2) audio recordings in English (European Parliament plenary speeches)and Spanish (European Parliament plenary speeches; Cortes Spanish Parliament speeches) developed by TALP[5] research group, distributed by ELDA and used in TC-START competitions for speech to speech translation. At other times, corpora is automatically obtained (for instance, English and Czech interview recordings of *Survivors of the Shoah Visual History Foundation* using in CL-SDR 2006[11] were transcribed using a ASR system with the consequent increase of transcription errors). There are other speech resources, recordings from telephone calls, dialogs, digits, short phrases, etc. but from the point of view of this work we are interested in spoken documents.

Focusing in European Spanish, the unique corpus with transcriptions that considers this language is the European Parliament and Cortes Spanish plenary speeches and other types of recordings are needed to test ASR systems for different kind of applications (for instance, voice queries in a Question Answering System over transcribed audio or video files) and domains (spoken documents concerning sports – broadcast sports news - or concerning international political issues - broadcast political news). In particular, this research work focuses on the use of TV broadcast contents to build valid test and training sets for ASR systems, mainly for Spanish. With this motivation, the research work introduced in this paper defines a platform for the evaluation of different ASR products (commercial or not) under the same conditions, i.e., using the same test collection and evaluation measures, and paying special attention to information retrieval applications. Moreover, a procedure to obtain literal transcription from audio resources is also defined in order to facilitate

---

[1] http://www.tcstar.org/
[2] http://trec.nist.gov/
[3] http://www.clef-campaign.org
[4] http://www.multimediaeval.org
[5] http://www.talp.cat

the creation of resources when there are not available literal transcriptions to test ASR systems. This is one of the goals covered in the BUSCAMEDIA[6] project, funded by the Spanish Ministry of Science and Innovation through the Centre for the Development of Industrial Technology (CDTI). This initiative is devoted to the study and development of advanced information retrieval, storage, generation and management mainly in Spanish, Catalan, Galician, Basque and English. BUSCAMEDIA searches for solutions to enhance multimedia information retrieval (IR) in the web. These solutions include the use of metadata related to the video or image or audio combined with content-based and text based retrieval techniques.

Current approaches concerning solutions that include ASR technologies are Google Voice[7] with the service Online Voicemail that gets transcribed messages delivered to mail inbox. Other vendors in the market, such as Autonomy Virage[8], include tools to perform audio and video indexing. To improve these applications, it is necessary to evaluate the accuracy of ASR technology before using it for information accessing applications.

In a research context there are several works, such as the work introduced in [2], a project to build a Spoken document retrieval system working on broadcast news repositories in Spanish and Basque. Viascribe[9] is a framework to do live subtitling in an educational environment. It uses de ViaVoice ASR system by IBM and offers to have different multimedia information sources integrated and synchronized. It permits to create a multimedia presentation integrating slides, captioning, videos, etc. Lecture Browser [9] is a web application developed by MIT to index and retrieve audio files proceeding from spoken lectures in the university.

More recently, APEINTA project [10] developed at Universidad Carlos III of Madrid has used ASR technology to overcome the barriers in the access to education and learning. In this inclusive proposal two mechanisms are used to overcome the communication barriers that still exist today in the classroom. One is the application of ASR mechanisms to provide real-time transcriptions, useful for all those students who have temporary or permanent hearing impairment. The other is the use of speech synthesis mechanisms to provide support for oral communication between teacher and students.

With the objective to investigate in techniques to characterize video and audio resources using transcriptions obtained using ASR systems, this paper answers the question *How to measure the performance of ASR technology in different contexts?* and it is focused on two aspects: (1) to propose a methodology that guide in evaluating an ASR system with a specific and suitable corpus and (2) how to define different scenarios of evaluation and how to prepare a corpus which serves as a gold-standard in a specific scenario.

The paper is organized as follows: section 2 introduces the methodology, section 3 describes how the methodology is used in a real evaluation using commercial ASR software and finally, section 4 shows several conclusions.

---

[6] http://www.cenitbuscamedia.es

[7] https://www.google.com/voice

[8] http://www.virage.com/rich-media/technology/index.htm

[9] http://liberatedlearningtechnology.com/wordpress/?page_id=509

2   Definition of a Methodology to Evaluate ASR Systems

Our final objective is to facilitate the evaluation process of ASR products to help us to select adequate software in a particular scenario that requires voice recognition. First of all a methodology to design and develop tests must be defined. This methodology is composed by the five steps included in Figure 1. Upper side shows the generic steps to follow and down side represents an instantiation of generic steps with the evaluation described in this paper with commercial ASR software.

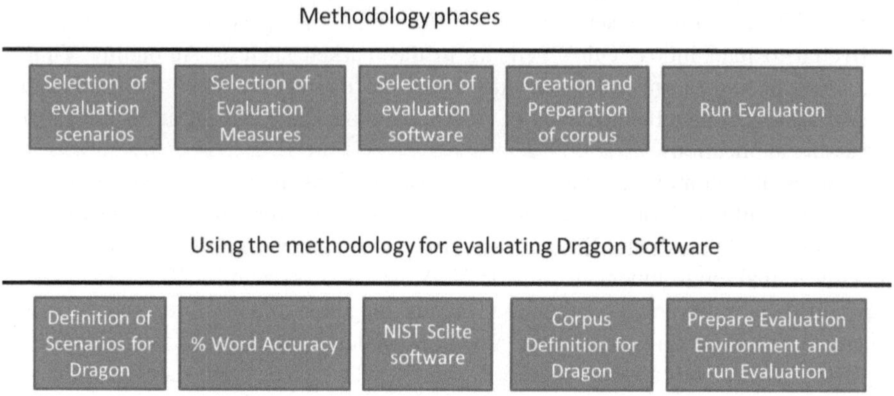

Methodology phases

| Selection of evaluation scenarios | Selection of Evaluation Measures | Selection of evaluation software | Creation and Preparation of corpus | Run Evaluation |

Using the methodology for evaluating Dragon Software

| Definition of Scenarios for Dragon | % Word Accuracy | NIST Sclite software | Corpus Definition for Dragon | Prepare Evaluation Environment and run Evaluation |

Fig. 1. Methodology phases with an example of use

The five steps that compose the methodology are:

1. Define and select the scenarios of evaluation: what are the contexts under the ASR system will work? For instance, if the ASR system is used with voice queries in a question answering system, if it is speaker-dependent, etc.
2. What will be measured? i.e., the confidence measures that indicate the performance of ASR system doing transcription of audio extracted from videos or other resources.
3. Selection of software evaluation: once the recognition is carried out, the performance according to confidence measures is evaluated. To do it in an appropriate way, specialized software is required.
4. Create and prepare a corpus: it is the most difficult step; depending on the scenario of evaluation as well as on the ASR system to be used, the corpus of speech recordings has to be carefully annotated.
5. Prepare evaluation environment and run evaluation to obtain the figures of confidence measures to evaluate performance.

Actually, these steps are not fully independent, there are relationships among them. For instance, the corpus preparation is influenced by the evaluation software to be used (the transcriptions of videos have to be formatted according to the required input in the evaluation system). In a similar way, the definition and selection of evaluation scenarios also affects corpus preparation. For example, if a scenario to test the performance of an ASR system with a specific speaker has to be defined, then the corpus has to contain enough video resources of this speaker.

2.1   First Step: Selection of Evaluation Scenarios

The central step in the methodology is to define the scenarios that will be used for evaluating. In this case, the parameters to be considered are: "domain" - which takes into account whether the domain of the audio (video) is focused on a specific matter or deals with general themes- ;"speaker"- that considers if there are one or several speakers in the audio (video)-;" training" –if the ASR system is going to be tested with no training, trained for a specific speaker or for several speakers - ; "test" that specifies the videos to be used in testing.

Using these characteristics seven resulting scenarios can be defined[10]:

(a)   Evaluation without training. In this scenario the ASR system is to be tested in initial conditions, that is, using the default acoustic and language models.

(b)   Evaluation with acoustic model training. In this scenario the acoustic model will be previously trained with audio resources from different speakers. This option is valid only for open source ASR software, such as Sphinx, where the acoustic model can be trained. In commercial systems, acoustic models can be adapted to a speaker but it is not possible to replace the model.

(c)   Evaluation with previous training. In this case a complete training of each ASR system (language model and acoustic model if possible) will be done using audio corpus (without selecting speakers).

(d)   Evaluation with speaker-oriented training. The ASR system is trained with a subset of the corpus in a speaker-dependent scenario (only one participant).

(e)   Evaluation with specific vocabularies. The ASR system is customized to work on a limited vocabulary previously defined.

(f)   Evaluation combining specific vocabulary and speaker dependence. In this scenario the ASR system will be trained and tested with videos of a specific subject of a specific speaker.

(g)   Evaluation without language model. Language models define the linguistic rules of a language and this knowledge allows the ASR system refines the options during recognition discarding linguistically invalid options and selecting those more grammatically appropriate. Without this language model the ASR exclusively depends on acoustic model (for instance, an ASR based on a unigram model to detect words using an HMM model where each word is represented by an automaton with transition between states representing phonemes, [12]).

2.2   Second Step: Selection of Evaluation Measures

To evaluate speech recognition systems, the output of the ASR system, called *hypothesis text*, is compared to a literal transcription of input audio, denoted as *reference text*. Standard measures used in speech recognition evaluation are [7]:

---

[10] Not all scenarios can be used in all ASR systems.

- *Word Error Rate*: it measures the percentage of incorrect words (*ps*-substitutions, *pi*- insertions, *pb*-eliminations) regarding the total number of words.

$$WER = \frac{ne}{pt} = \frac{ps + pi + pb}{pt}$$

where *ne* is the total number of errors in hypothesis text and *pt* is the number of total words in the reference text.

- *Word Accuracy*: it measures the total number of correct words regarding the total number of words.

$$WAcc = 1 - WER = \frac{pc}{pt}$$

where *pc* is the total number of correct words in hypothesis text.

Apart from these quantitative measures there are qualitative measures that allow understanding bad results in word accuracy. A previous work, [8], defined a visual framework whose objective was to do a qualitative evaluation of speech transcriptions generated by an ASR system apart from obtaining word accuracy and word error rate. Mainly we are interested in analyzing the type of errors concerning *Out Of Vocabulary words* which are not included in the ASR dictionary, taking into account their grammatical categories (named entities, nouns, verbs, etc.).These qualitative measures do not have to do with user satisfaction using the result of recognition process.

Due to ASR technology is error prone, concerning the user satisfaction measures, there are several works that investigate in providing corrections to the ASR output in user interfaces in different applications. For instance, if the ASR technology is applied to voice queries that are the input to a search engine, it would be adequate to test the user satisfaction with the query transcription considering that not all the words are equally important (a search engine could retrieve relevant documents given a query with named entities right recognized but if named entities are incorrectly recognized the search engine could not retrieve relevant documents). In this line, [14] and [15] proposed solutions to provide alternatives to wrong words in speech input interfaces.

2.3   Third Step: Selection of Evaluation Software

After defining the measurements that are going to be used to evaluate the system, next step selects the evaluation software to test the quality of recognition process. A well-known software to evaluate speech recognition is Sclite [6] that is part of the Scoring Toolkit developed (SCTK) developed by NIST (National Institute of Standards and Technologies). The goal of Sclite is to evaluate an ASR system by comparing a manual transcription with the automatic transcription obtained from the ASR. To obtain this comparison the Sclite tool needs two files: a *Reference File* containing the manual transcription obtained by an expert and a *Hypothesis File* containing the automatic transcription returned by ASR.

Both the reference and hypothesis file can take different formats but we have preferred using the sentence time marked STM format for the reference file and the word time marked CTM for the hypothesis file. The reason to use these formats is that the result coming from Sclite is better as long as the time alignment is used during the matching process.

2.4 Fourth Step: Create and Prepare a Corpus

The videos/audio resources have to be collected and classified according to different parameters: audio format, domain, speakers, noise, music and other characteristics that should have correspondence with the scenarios defined in the second step. Moreover the corpus has to be divided in training and testing parts depending on the evaluation scenario, in order to perform a cross-validation evaluation.

3   Some Experiments Applying the Methodology to a Commercial ASR System

To accomplish the fourth step a video collection to test ASR systems was prepared. It is composed of 15 generic TV Broadcast news (with duration of one hour each), 10 videos about sport news videos and 10 videos containing weather forecasts (approx., 10 minutes each). These resources had to be split in segments of approx. 10 minutes due to (a) allowing configuring different training-testing parts (b) software limitations both in ASR system and in NIST Score Toolkit evaluation software.

The Spanish TV Broadcast news videos contain a main newsreader and some secondary newsreaders (weather, sports, etc). There are also many live connections inserted in the news reading to make interviews or reports. Each external connection is characterized by different speakers and noisy environment. They deal with generic subjects.

Their transcription is stored in an 'standard' XML file dividing the transcription into sentences and containing each sentence the initial and final time marks (in seconds), a speaker identification and the transcription of the sentence. The sentences are delimited by a long silence in the speaker's speech. Each weather forecast contains one speaker and has a noiseless environment.

To perform the last step (prepare evaluation environment and run evaluation) on the DNS software, three scenarios described in section 3 have been selected (a, c and d). DNS provides two manners to train a speaker model, one is using the commercial version and other is using different functions that are provided by Dragon SDK (we have used the second option by implementing a program which receives as input an audio file with its corresponding transcription in a raw text file). Four different trainings were defined:

1. Evaluation without training: using the default acoustic and language model provide by DNS (*scenario a*).
2. Evaluation with previous speaker independent training (*scenario c*).
3. Evaluation with specific vocabularies (*scenario e*)
4. Evaluation combining specific vocabulary and speaker dependent training (*scenario f*)

Table 1. Speech Corpus features

	N° of segments	Duration / segment	Source	Speakers/ segment	% Noise aprox	% Music aprox	% Overlap ping Voices aprox
TV Broadcast News (in Spanish)	10	9 min aprox.	RTVE	10-15 aprox.	62 %	2 %	2 %
Weather Forecasts (in Spanish)	13	10 min aprox.	RTVE	1	*	5 %	0 %
*All segments with background music							

The three last trainings were decomposed in two sub-scenarios due to DNS facility to train the user model using audio files and their corresponding transcription (notice that the DNS Spanish model has been used). So, we distinguish among: *Short enrollment*, where DNS was trained using one video with a length of, approximately, 10 minutes; and *Long enrollment*, where the DNS user model was trained using 7 videos with a mean length of 10 minutes).

For long enrollment experiments, seven speaker models were created and trained, which were tested using a ten minutes video randomly selected from the corpus/collection (the video used for test is not the same used for training).

Table 2 shows the experiments that have been completely developed and evaluated. Word accuracy values are very similar in the three cases. We believe that training using video segments where 10/12 different speakers are taking part, with noise, music and overlapping voices is not a good material to train user models.

Initial test runs showed that some settings are required in the corpus preparation phase. Some of them are: (1) different encodings appearing during execution, i.e., DNS returns the output encoded as "ISO-LATIN-1" while Sclite accepts "ANSI" encodings and manual transcriptions are "UTF-8" encoded. (2) errors concerning treatment of numbers (in manual transcription files numbers are written using figures while ASR recognizes them as alphanumeric characters), punctuation marks (no ASR system obtains transcriptions with punctuation marks but reference manual transcriptions used to test the ASR systems contain them). Several human transcription errors are unavoidable but they require to define a semi-automatic process that helps annotators to do a quality transcription free of errors (apart from lexical errors, errors mainly related to temporal synchronization among output from DNS an transcription segments are also frequent in manual transcriptions). Finally, specific problems of DNS recognition, such as enclitic pronouns which are separated by DNS system (leading to matching errors in the evaluation phase), must also be taken into account.

Table 2. Preliminary results using DNS system

	Scenario (a) Without Enrollment	Scenario (c) with Short Enrollment	Scenario (c) with Long Enrollment
% Correct	68,8%	69,8%	71,7%
% Substitutions	15,8%	14,3%	13,8%
% Deletions	15,4%	15,9%	14,5%
% Insertions	3,8%	3,3%	3,9%
% Word Accuracy	64,9%	66,5%	67,8%

4   Some Conclusions

The work accomplished up to now has allowed us to face different problems that have to be fixed previously to ASR system testing. As an example, the first experiment we ran, evaluated scenario c with a short training in DNS using the same video fragment in training and test. The hypothesis was that word accuracy should be near 100% but surprisingly the result was near 85%. This result means that we have to be extremely careful in creating and annotating the corpus and understanding the internal processing in the ASR system. Specific options in these systems to deal with characters, punctuation marks and time segmentation, must be deeply studied in order to have a powerful test bed.

Video transcription generation for our corpus is based on two methods. Initially, Aegisub Software[11] for subtitling was used to segment TV Broadcasts news videos in fragments of 10 minutes with their corresponding phrases with temporal marks, as required at the input in Sclite to evaluate output recognition. Unfortunately, using this software was not a good idea because it is too difficult to be precise during manual segmentation. As an alternative method, we have decided to use the proper DNS to detect the duration of segments to be considered in the corpus. This helps annotators to do quality transcriptions.

First accuracy figures shown in Table 2 should be taken as preliminary results, showing an almost negligible accuracy increase comparing trained and no trained experiments. Future lines of work will be centered on the study of the results of the evaluation, assuring that problems drawn in Section 7 are surpassed, and performing the rest of experiments in order to evaluate the amount of training needed to get good

---

[11] http://www.aegisub.org/

quality transcriptions. Then, DNS will be changed to different ASR products, such as Sphynx or Windows Speech Recognition.

Acknowledgements. This work has been partially supported by the Spanish Center for Industry Technological Development (CDTI, Ministry of Industry, Tourism and Trade), through the BUSCAMEDIA Project (CEN-20091026). Authors would like to thank all BUSCAMEDIA partners for their knowledge and contribution and also by MA2VICMR: Improving the access, analysis and visibility of the multilingual and multimedia information in web for the Region of Madrid (S2009/TIC-1542).

References

[1] Baeza-Yates, R., Ribeiro-Neto, B.: Modern Information Retrieval: The Concepts and Technology behind Search, 2nd edn. ACM Press Books (2011)

[2] Varona, A., Rodríguez Fuentes, L.J., Penagarikano, M., Nieto, S., Diez, M., Bordel, G.: Search and access to information contained in the speech of multimedia resources. Procesamiento del Lenguaje Natural 45, 317–318 (2010)

[3] Sphinx, http://cmusphinx.sourceforge.net/

[4] The HTK Speech Recognition Toolkit, http://htk.eng.cam.ac.uk/

[5] Dragon, http://www.nuance.com/naturallyspeaking/

[6] Sclite, ftp://jaguar.ncsl.nist.gov/current_docs/sctk/doc/sclite.htm

[7] Dybkjaer, L., Hemsen, H., Minker, W.: Evaluation of Text and Speech systems, pp. 1–64, 99–124. Springer (2007)

[8] Moreno, J., Garrote, M., Martínez, P., Martínez-Fernández, J.L.: Some experiments in evaluating ASR systems applied to multimedia retrieval. In: Detyniecki, M., García-Serrano, A., Nürnberger, A. (eds.) AMR 2009. LNCS, vol. 6535, pp. 12–23. Springer, Heidelberg (2011)

[9] Spoken lecture processing system, MIT, http://web.sls.csail.mit.edu/lectures/

[10] Iglesias, A., Moreno, L., Ruiz-Mezcua, B., Pajares, J.L., Jiménez, J., López, J.F., Revuelta, P., Hernández, J.: Web Educational Services for All: The APEINTA project, Web Accessibility Challenge. In: 8th International Cross-Disciplinary Conference on Web Accessibility, Hyderabad, India (2011)

[11] Oard, D., Wang, J., Jones, G., White, R., Pecina, P., Soergel, D., Huang, X., Shafran, I.: Overview of the CLEF-2006 Cross-Language speech retrieval track. In: Peters, C., Clough, P., Gey, F.C., Karlgren, J., Magnini, B., Oard, D.W., de Rijke, M., Stempfhuber, M. (eds.) CLEF 2006. LNCS, vol. 4730, pp. 744–758. Springer, Heidelberg (2007)

[12] Huang, X., Jack, M., Ariki, Y.: Hidden Markov Models for Speech Recognition. Edinburgh University Press (1990)

[13] De Mori, R., Bechet, F., Hakkani-Tur, D., McTear, M., Riccardi, G., Tur, G.: Spoken Language Understanding: A Survey. IEEE Signal Processing Magazine 25, 50–58 (2008)

[14] Ogata, J., Goto, M.: Speech repair: quick error correction just by using selection operation for speech input interfaces. In: Proc. Eurospeech 2005, pp. 133–136 (2005)

[15] Sarma, A., Palmer, D.: Context-based speech recognition error detection and correction. In: Proceedings of HLT-NAACL (2004)

Context-Aware Features for Singing Voice Detection in Polyphonic Music

Vishweshwara Rao, Chitralekha Gupta, and Preeti Rao

Department of Electrical Engineering
IIT Bombay, Mumbai - 76, India
{vishu,chitralekha,prao}@ee.iitb.ac.in

Abstract. The effectiveness of audio content analysis for music retrieval may be enhanced by the use of available metadata. In the present work, observed differences in singing style and instrumentation across genres are used to adapt acoustic features for the singing voice detection task. Timbral descriptors traditionally used to discriminate singing voice from accompanying instruments are complemented by new features representing the temporal dynamics of source pitch and timbre. A method to isolate the dominant source spectrum serves to increase the robustness of the extracted features in the context of polyphonic audio. While demonstrating the effectiveness of combining static and dynamic features, experiments on a culturally diverse music database clearly indicate the value of adapting feature sets to genre-specific acoustic characteristics. Thus commonly available metadata, such as genre, can be useful in the front-end of an MIR system.

1 Introduction

The automatic identification of audio segments within a song, that contain the singing voice (vocal part) is important in several Music Information Retrieval (MIR) applications such as artist identification [1], voice separation [2] and lyrics alignment [3]. SVD is typically viewed as an audio classification problem where features that distinguish vocal segments from purely instrumental segments in music are fed to a machine-learning algorithm previously trained on manually labeled data. Until recently, singing voice detection algorithms employed solely static features, typically comprising frame-level spectral measurements such as combinations of mel-frequency cepstral coefficients (MFCCs) [1], [2], [4]-[6], warped or perceptually derived linear predictive coefficients (LPCs) [1], [7]-[9], log frequency power coefficients (LFPC) [10], harmonicity related features [11]-[13] and other spectral features such as flux, centroid and roll-off [14], [15].

A consideration of acoustic attributes necessary for the detection of vocal segments in music fragments by humans is interesting for its potential in guiding the search for suitable features for the task. An experiment in which subjects listened to short excerpts (less than 500 ms long) of music from across diverse genres showed that human listeners can reliably detect the presence of vocal

M. Detyniecki et al. (Eds.): AMR 2011, LNCS 7836, pp. 43–57, 2013.
© Springer-Verlag Berlin Heidelberg 2013

content in such brief excerpts [16]. The presence of note transitions in the excerpt was found to be especially useful, indicating that both static features and dynamic features (changes) provide important perceptual cues to the presence of vocalization.

Dynamic features for singing voice detection have largely been confined to feature derivatives, representing very short-time dynamics. A few studies have explored the observed pitch instability of the voice relative to most accompanying instruments in the form of features representing longer-term dynamics of pitch and/or the harmonics such as arising from vibrato and tremolo in singing [10], [17]. Clearly, there is scope for improvement in terms of capturing the essential differences in the dynamics of the singing voice and musical instruments in a compact and effective way.

While most research results in MIR are reported on collections drawn from one or another culture (mostly Western), we are especially interested in features that work cross-culturally. It would be expected that certain feature attributes are more discriminative on particular music collections than on others, depending on the musical content due to the inherent diversity of both singing styles and instrumentation textures across cultures [18]. The work reported in this paper is an attempt to study this. A similar approach underlies the work on collection-specific features for image retrieval, also appearing in this volume [19].

We choose to focus on 'vocal' music i.e. where the singing voice is the predominant melodic entity whenever present. Further, as far as possible, the instrumental sections of the song include a predominant melodic instrument. One of the categories that was badly classified, and also negatively influenced the training set effectiveness, in the study of [20] corresponded indeed to songs with predominant melodic instruments or with singing co-occurring with such instruments. Paralleling this observation, are studies on predominant musical instrument identification in polyphony which state that pitched instruments are particularly difficult to classify due to their sparse spectra [21]. Thus our choice of evaluation datasets is guided by the known difficulty of the musical context as well as the wide availability of such a category of music cross-culturally. We consider the effective extraction and evaluation of static and dynamic features on a dataset of vocal music drawn from Western popular, Greek and three distinct Indian genres: North Indian (Hindustani) classical, South Indian (Carnatic) and popular (Bollywood or film music).

2   Dominant Source Spectrum Isolation

Previous studies on SVD have extracted features by processing the short-time Fourier transform (STFT) without attempting to isolate the spectra of individual sound sources. Recently, a spectral processing technique called accompaniment sound reduction, which uses the predominant-F0, for robust singing voice modeling in polyphony was presented by Fujihara et.al.[22]. In this section we describe a method for dominant source spectrum isolation along the lines of their method but with some important enhancements.

Harmonic sinusoidal modeling is applied to detect and track F0-specific harmonic components from the spectrum. We use a predominant F0 extraction system designed for robustness in the difficult context of concurrent pitched accompaniment [23]. In order to study the comparative performance of features unobscured by possible pitch detection errors, we carry out feature extraction in both automatic and semi-automatic modes of pitch detection for the dominant source spectrum isolation [24]. In the latter mode, pitch analysis parameters are selected considering a priori information on the pitch range of the voice in the given piece of music.

The audio signal is processed at a frame rate of 100 frames/sec with the STFT computed from 40 ms long Hamming windowed analysis frames. Rather than identify spectral local maxima in the vicinity of expected harmonic locations as detected harmonics, as done in [22], we first extract reliable sinusoidal partials using a main-lobe matching technique, which we have previously found to be particularly robust to polyphony and signal non-stationarity [25]. We then search the sinusoidal space in a 50-cent neighborhood (subject to a 50 Hz absolute limit) of expected harmonic locations, based on known predominant F0, to identify local harmonic components. This results in a harmonic line spectrum for each analysis frame.

Next, partial tracking is applied to the harmonic line spectra over time for better source spectral isolation during transients and note changes. Similar to Serra [26], partial tracking is improved by biasing trajectory formation towards expected harmonic locations based on the detected pitch. Tracks are formed and ordered by their respective harmonic index. We apply a one semitone threshold on track continuation i.e. a track will 'die' if there does not exist any sinusoid within 1 semitone of the last tracked frequency. In addition to frequency proximity, we incorporate sinusoid amplitudes in the cost function, as given below, to account for high amplitude partials that might lose out to spurious nearby peaks.

$$J = \left| (\omega_n^k - \omega_{n-1}^m) \times \log(A_n^k / A_{n-1}^m) \right| \tag{1}$$

where ω_{n-1}^m and A_{n-1}^m are the frequency and amplitude resp. of the mth harmonic track in frame $n-1$ and ω_n^k and ω_{n-1}^m are the frequency and amplitude resp. of the kth local sinusoid in the frame n that is competing for joining the track.

Rather than extract features from the discrete frequency spectra obtained after harmonic sinusoidal modeling, we use smooth spectra derived by the log-linear interpolation of the harmonic spectral amplitudes as follows. Given a set of estimated amplitudes S(ω_1), S(ω_2), , S(ω_L) at L uniformly spaced harmonic frequencies ω_1, ω_2,, ω_L we generate spectral amplitudes Q(Θ) at fixed DFT bin spacing as shown below:

$$Q(\Theta_j) = 10^{\log|S(\omega_k)| + (\frac{\Theta_j - \omega_k}{\omega_{k+1} - \omega_k})(\log|S(\omega_{k+1})| - \log|S(\omega_k)|)} \tag{2}$$

where $\omega_k < \Theta_j < \omega_{k+1}$. The interpolation serves to make any spectral envelope features extracted subsequently less dependent on the F0 and thus more representative of source timbre.

3   Feature Extraction

In this section we describe the different static and dynamic features investigated, motivated by the signal characteristics. Static features are computed locally over short sliding analysis frames while dynamic features are computed over larger non-overlapping time windows called texture windows. To avoid confusion, we will use the term *frame* to refer to the analysis frame and *window* for the texture windows. Silence frames, i.e. frames with energy lower than the song-level global maximum energy by more than 30 dB, are not processed. The features for such frames are interpolated from valid feature values in adjacent frames.

3.1   Static Timbral Features

In previous work on SVD the most commonly used features are static timbral descriptors, which attempt to capture the timbral aspects of musical sounds and do not utilize any specific long-term traits of the singing voice or instruments. Rocomora and Herrera compared the performance of a comprehensive list of static features in an SVD task and found that MFCCs resulted in the best classification accuracies [27]. The tested features were however all extracted from the overall polyphonic spectrum.

Recently Fujihara and Goto proposed that the feature set of F0 and harmonic powers were found to be highly discriminatory for SVD when source spectral isolation was possible [3]. The harmonic powers were normalized for each song. The normalized power of the hth component at time t is given by

$$p_h^{'t} = \log p_h^t - \frac{\sum_t \sum_h \log p_h^t}{T \times H} \qquad (3)$$

where $p_h^{'t}$ represents the original power, T is the total number of frames in the song, and H is the number of harmonic components considered per frame.

The interpolated spectrum of Eq. (2) provides a spectral envelope that is potentially indicative of the underlying instrument's resonances or formants. Two more features: the sub-band spectral centroid (SC) and sub-band energy (SE) [28] as given by

$$SC = \frac{\sum_{k=k_{low}}^{k_{high}} f(k)|X(k)|}{\sum_{k=k_{low}}^{k_{high}} |X(k)|} \qquad SE = \sum_{k=k_{low}}^{k_{high}} |X(k)|^2 \qquad (4)$$

where $f(k)$ and $|X(k)|$ are frequency and magnitude spectral value of the kth frequency bin, and k_{low} and k_{high} are the nearest frequency bins to the lower and upper frequency limits on the sub-band respectively. SE is normalized by its maximum value for a song. The sub-band for SC ranges from 1.2-4.5 kHz and that for SE is from 300 to 900 Hz.

3.2   Dynamic Timbral Features (C2)

The explicit modeling of temporal dynamics, as an important component of perceived timbre, has found a place in recent research on musical instrument recognition [29], [30]. Based on similar considerations, features linked to the temporal evolution of the spectral envelope can be designed to capture specific attributes of the instrument sound. For example, in view of the well-known source-filter model of sound production applicable to human speech and to musical instruments, variations in spectral envelope with time represent variations of the filter component independent of the source component (i.e. F0). Such a dynamic feature could potentially discriminate between singing voice and (even similarly pitch-modulated expressive) musical instruments due to the absence of formant articulation dynamics in the latter. Another distinguishing aspect could be the attack-decay envelope peculiar to a particular instrument as reflected in the variation of spectral amplitudes over note durations.

One of the problems with the effective use of timbral dynamics for instrument classification in polyphony was found to be the lack of ability to pay selective attention to isolated instrumental spectra [31]. Here we describe the extraction of the dynamics of the two static timbral features using our isolated dominant source spectral representation. The first of these is the sub-band spectral centroid (SC) as computed in the previous section. The specific sub-band chosen [1.2-4.5 kHz] is expected to enhance the variations in the 2nd, 3rd and 4th formants across phone transitions in the singing voice. This feature is expected to remain relatively invariant over note transitions in the instrument. Although the band-limits are restrictive, even very high pitched instruments will have at least 3 harmonics present within this band so that their spectral envelope can be relatively well represented. The second feature for dynamic analysis is the sub-band energy (SE) as computed in the previous section. The band limits for SE ([300-900 Hz]) are expected to enhance the fluctuations between voiced and unvoiced utterances while remaining relatively invariant to small instrumental note transitions. Fluctuations in this feature should be evident even if the signal representation captures some pitched accompanying instrument information during unvoiced sung sounds.

To capture meaningful temporal variations in the dynamics of the above timbral features, it is necessary to choose the duration of the observation interval appropriately [29]. We choose three different time scales (texture windows) for our feature set: 0.5 sec (long note duration), 1 and 2 sec intervals (to capture note articulation changes in both fast and slow singing). We represent the dynamics via the standard deviation (std. dev.) and specific modulation energies over the different observation intervals. These modulation energies are represented by a modulation energy ratio (MER). The MER is extracted by computing the DFT of the feature trajectory over a texture window and then computing the ratio of the energy in the 1-6 Hz region in this modulation spectrum to that in the 1-20 Hz region as shown below:

$$MER = \frac{\sum_{k=k_{1Hz}}^{k_{6Hz}} |Z(k)|^2}{\sum_{k=k_{1Hz}}^{k_{20Hz}} |Z(k)|^2} \tag{5}$$

where $Z(k)$ is the DFT of the mean-subtracted feature trajectory $z(n)$ and k_{fHz} is the frequency bin closest to fHz. We assume that the fastest syllabic rate possible, if we link each uttered phone to a different note in normal singing, should not exceed 6 Hz. Steady note durations are not expected to cross 2 seconds. The std. dev. and MER of the above features are expected to be higher for singing than instrumentation.

3.3 Dynamic F0-Harmonic Features (C3)

Singing differs from several musical instruments in its expressivity, which is physically manifested as the instability of its pitch contour. In western singing, especially operatic singing, voice pitch instability is marked by the widespread use of vibrato, a periodic, sinusoidal modulation of phonation frequency during sustained notes [32]. Within non-western forms of music, such as Greek Rembetiko and Indian classical music, voice pitch inflections and ornamentation are extensively used as they serve important aesthetic and musicological functions. On the other hand, the pitch contours of several accompanying musical instruments, especially keyed instruments, are usually very stable and incapable of producing pitch modulation.

There has been limited previous work on utilizing voice-pitch instability to SVD. Shenoy, Wu and Wang [33] exploit pitch instability in an indirect way by applying a bank of inverse comb filters to suppress the spectral content (harmonics) of stable-pitch instruments. Nwe and Li [10] made use of a bank of band-pass filters to explicitly capture the extent of vibrato within individual harmonics upto 16 kHz. Regnier and Peeters attempted a more direct use of frequency and amplitude instability of voice harmonics (manifested as vibrato and tremolo respectively in western music) [17]. Their method is based on the observation that the extents of vibrato and tremolo in singing are different than those for most instruments. The authors have also previously used pitch stability cues to prune flat-pitched instruments tracks in a harmonic sinusoidal model [34].

In this study we do not restrict ourselves to targeting particular types of pitch modulation such as vibrato but extract some statistical descriptors (mean, median, std. dev.) of general pitch instability-based features over texture windows of expected minimum note duration (here 200 ms). These features are the first-order differences of the predominant-F0 contour and the subsequently formed harmonic frequency tracks. The track frequencies are first normalized by harmonic index and then converted to the logarithmic cents scale so as to maintain the same range of variation across harmonics and singers' pitch ranges. For the latter we group the tracks by harmonic index (harmonics 1-5, harmonics 6-10, harmonics 1-10) and also by low and high frequency bands ([0-2 kHz] and [2-5 kHz]). This separation of lower and higher harmonics/frequency bands is due to the observation that when the voice pitch is quite stable, the lower harmonics do not display much instability but this is clearly visible in the higher harmonics. However when the voice pitch exhibits large modulations the instability in the lower harmonic tracks is much more clearly observed but often the higher harmonic tracks are distorted and broken because of the inability of the sinusoidal model to reliably track their proportionately larger fluctuations. We also

Table 1. List of features in each category. Bold indicate finally selected feature.

C1 Static timbral	C2 Dynamic timbral	C3 Dynamic F0-Harmonic
F0	Δ 10 Harmonic powers	Mean and **median** of Δ**F0**
10 Harmonic powers	ΔSC and ΔSE	Mean, median and Std.Dev. of ΔHarmonic $\epsilon[0\ 2\ \text{kHz}]$
Spectral centroid (SE)	**Std. Dev.** of **SC** for 0.5, 1 and 2 sec	Mean, median and Std.Dev. of ΔHarmonic $\epsilon[2\ 5\ \text{kHz}]$
Sub-band energy (SE)	**MER** of **SC** for 0.5, 1 and **2** sec	Mean, median and Std.Dev. of ΔHarmonics 1 to 5
	Std. Dev. of **SE** for **0.5**, 1 and 2 sec	Mean, **median** and Std.Dev. of Δ**Harmonics 6 to 10**
	MER of SE for 0.5, 1 and 2 sec	Mean, **median** and Std.Dev. of Δ**Harmonics 1 to10**
		Ratio of mean, median and Std.dev. of ΔHarmonics 1 to 5 : ΔHarmonics 6 to 10

compute the ratio of the statistics of the lower harmonic tracks to those of the higher harmonic tracks since we expect these to be much less than 1 for the voice but nearly equal to 1 for flat-note instruments.

A complete list of features appears in Table 1. All features are brought to the time-scale of 200 ms long decision windows. The frame-level static timbral features, generated every 10 ms, are averaged over this time-scale and the timbral dynamic features,generated over larger windows: 0.5, 1 and 2 sec, are repeated within 200 ms intervals. The F0-harmonic dynamic features were generated at 200 ms non-overlapping windows in the first place and do not need to be adjusted. Next, feature subset selection is applied to identify a small number of highly predictive features and remove as much redundant information as possible. Reducing the dimensionality of the data allows machine learning algorithms to operate more effectively from available training data. Feature selection is achieved by measuring the information gain ratio of the feature with respect to a class [35]. Each feature is assigned a score based on the information gain ratio to obtain a ranked feature list.

4   Classification

Statistical classification methods are very effective in exploiting the overall information provided about the underlying class by the set of diverse features if suitable data is provided for the training of the statistical models. Previous studies on SVD have employed a variety of classifiers. We use a standard GMM classifier [36] for the evaluation of our features. Using 4 mixtures for each of the two models (singing voice and instrumentation) with full covariance matrices was found suitable in preliminary testing. The dimensionality of the GMM is determined by that of the feature vector. We would like to study the effectiveness of individual feature sets as well as that of the combination of individual classifiers trained with individual feature sets, which can improve the effectiveness of the system while offsetting difficulties arising from high dimensionality [37]. Combining the likelihood scores of classifiers is particularly beneficial if the corresponding individual feature sets represent complementary information about the underlying signal. Weighted linear combination of likelihoods provides a flexible method of combining multiple classifiers with the provision of varying the weights to optimize performance [37].

5   Experiments

The singing voice detection task is carried out on a database comprising excerpts from 5 distinct music genres. A feature set of the first 13 MFCCs extracted from the frame-level magnitude spectrum applied to a GMM classifier with 4 mixtures per class is considered as a baseline system. As mentioned earlier, Rocamora and Herrera had found the performance of the MFCC features among the best performing for in the SVD task [27]. The newly proposed features of the present work are applied to the same classifier framework in order to evaluate the performance improvement with respect to the baseline feature set and to derive a system based on possible feature combinations that performs best for a specific genre and across genres.

5.1   Database Description

All the audio excerpts in our database contain polyphonic music with lead vocals and dominant pitched melodic accompaniment, and are in 22 kHz 16-bit mono format. Vocal and purely instrumental sections of songs categorized as above from five different genres of music have been selected - Western popular, Greek Rembetiko, Indian popular (Bollywood), North Indian classical (Hindustani) and South Indian classical music (Carnatic). The Western and Greek clips have been selected from the datasets used in [9] and [38] respectively. The total size of the database is about 65 min. with roughly 13 min. per genre. Information about the number of songs, and vocal and instrumental durations for each genre appears in Table 2. In a given genre a particular artist is represented by only one song.

The selected genres are marked by distinct singing styles and instrumentation. A noticeable difference between the singing styles of the Western and non-Western genres is the extensive use of pitch-modulation (other than vibrato) in the latter. Pitch modulations further show large variations across non-Western

Table 2. Duration Information of Audio Database

Genre	Number of songs	Vocal duration	Instrumental duration	Overall duration
I. Western	11	7m 19s	7m 02s	14m 21s
II. Greek	10	6m 30s	6m 29s	12m 59s
III. Bollywood	13	6m 10s	6m 26s	12m 36s
IV. Hindustani	8	7m 10s	5m 24s	12m 54s
V. Carnatic	12	6m 15s	5m 58s	12m 13s
Total	**45**	**33m 44s**	**31m 19s**	**65m 03s**

genres in the nature, shape, extents, rates and frequency of use of specific pitch ornaments. Further, whereas Western, Greek and Bollywood songs use syllabic singing with meaningful lyrics, the Hindustani and Carnatic music data is dominated by melismatic singing (several notes on a single syllable in the form of continuous pitch variation). The instruments in Indian popular and Carnatic genres are typically pitch-continuous such as the violin, saxophone, flute, *shehnai*, and *been*, whose expressiveness resembles that of the singing voice in terms of similar large and continuous pitch movements. Although there are instances of pitch-continuous instruments such as electric guitar and violin in the Western and Greek genres as well, these, and the Hindustani genre, are largely dominated by discrete-pitch instruments such as the piano and guitar, accordion and the harmonium. A summary of genre-specific singing voice and instrumental characteristics appears in Table 3.

5.2 Selected Features

Each of the feature sets (C1, C2 and C3) is fed to the feature selection system to generate a ranked list for each genre. A feature vector comprising the top-N features common across genres was tested for SVD in a cross-validation classification experiment to select N best features. For C1 it was observed that using all the features in this category consistently maximized the classification accuracies across genres and so we did not discard any of these features. For C2 and C3 we observed that the top six selected features for each of the genres consistently maximized their respective classification accuracies. The finally selected features in each of the categories appear in bold in Table 1.

In the dynamic timbral feature set, the Δ values of the static features are ignored by the feature selection algorithm in favour of the std. dev. and MER values of the SC and SE. The feature selection algorithm took into account the expected high degree of correlation between the same dynamic features at different time-scales and only selected at most one time-scale for each dynamic feature. For the F0-harmonic dynamic feature set, the final selected features (C3) are the medians of ΔF0 and ΔHarmonic-tracks rather than their means or std. dev. The choice of medians was seen to be driven by the common occurrence of intra-window flat-pitched instruments note-transitions where the F0/Harmonic

Table 3. Genre-specific singing and instrumental characteristics

Genre	Singing	Dominant Instrument
I. Western	Syllabic. No large pitch modulations. Voice often softer than instrument.	Mainly flat-note (piano, guitar). Pitch range overlapping with voice.
II. Greek	Syllabic. Replete with fast, pitch modulations.	Equal occurrence of flat-note plucked-string /accordion and of pitch-modulated violin.
III. Bollywood	Syllabic. More pitch modulations than Western but less than other Indian.	Mainly pitch-modulated wood-wind & bowed instruments. Pitches often much higher than voice.
IV. Hindustani	Syllabic and melismatic. Varies from long, pitch-flat, vowel-only notes to large & rapid pitch modulations.	Mainly flat-note harmonium (woodwind). Pitch range overlapping with voice.
V. Carnatic	Syllabic and melismatic. Replete with rapid pitch modulations.	Mainly pitch-modulated violin. F0 range generally higher than voice but has some overlap in pitch range.

tracks make a discontinuous jump. In such cases, the means and standard deviations of the Δs exhibit large values as opposed to the relatively unaffected median values which remain low.

5.3   Evaluation

An N-fold cross-validation classification experiment is carried out within each genre. Since the durations of different songs within a particular genre are unequal we consider each song to be a fold so as to avoid the presence of tokens of the same song in the training and testing data to achieve a 'Leave 1 Song out' cross-validation. This is done separately for the semi- and fully-automatic F0 extraction based source spectrum isolation methods. In each case, we first evaluate the performance of the baseline features (MFCCs), before and after applying dominant source spectrum isolation. We next evaluate the performance of the different categories of feature sets individually (C1, C2 and C3). Further, we evaluate the performance of different feature set combinations: C1+C2, C1+C3 and C1+C2+C3, using a linear combination (equal weights) of the log-likelihood outputs per class of separate classifiers for each feature set. Vocal/non-vocal decision labels are generated for every 200 ms texture window. While the ground-truth labels for the Western and Greek genres were available with the datasets, the

Table 4. %Correct classification for different genres using semi-automatic predominant F0 extraction;Bold indicates best achieved in each genre

Genre	I	II	III	IV	V	Total
Baseline	77.2	66.0	65.6	82.6	83.2	74.9
MFCCs(after source isolation)	78.9	77.8	78.0	85.9	85.9	81.2
C1	79.6	77.4	79.3	82.3	87.1	81.0
C2	72.0	77.9	80.0	70.1	65.3	73.2
C3	64.3	77.0	68.3	83.7	70.2	72.6
C1+C2	**82.3**	83.6	85.4	83.3	86.8	84.2
C1+C3	80.2	83.4	81.7	**89.7**	**88.2**	84.5
C1+C2+C3	81.1	**86.9**	**86.4**	88.5	87.3	**85.9**

remaining datasets were manually labeled. In all cases classification performance is given by the percentage of decision windows that are correctly classified [2].

5.4   Results and Discussion

The 'Leave 1 Song out' cross-genre classification results for semi- and fully-automatic predominant F0 extraction based source spectrum isolation methods are given in Tables 4 and 5 respectively. From Table 4, we see that the best over-all performance is achieved for the combination of all three feature sets and is significantly (10-12%) higher than the baseline performance. For the static feature comparison it can be seen that the feature sets C1 and MFCCs after source isolation show similar performance and are, in general, superior to the baseline features(non-source-isolated MFCC). The clear superiority of the C1+C2+C3 feature combination over the static feature set C1 and over the MFCC feature set after source isolation can also be observed by the across-genre average vocal precision v/s recall curves in Fig. 1. A detailed analysis of the genre-specific performance of each feature set follows. The feature set C2 shows relatively high performance for the Western, Greek and Bollywood genres as compared to the Hindustani and Carnatic genres. This can be attributed to the presence of normal syllabic singing in the former and long duration vowel and melismatic singing in the latter. The relatively high performance of this feature set in the Bollywood genres where the instruments are mainly pitch-continuous corrobo-rates with the static timbral characteristics of these instruments despite their continuously changing pitch. Table 5 shows reduced results for different feature sets and their combinations as compared to Table 4. This is due to the pitch detection errors inherent in the fully-automatic F0 extraction system. However the general trends across different feature sets hold for this case as well, with the results of Table 4 providing an upper limit on performance achievable with a better pitch detector.

The suitability of C2 and C3 for specific signal conditions can be understood from Fig. 2a and 2b, which show spectrograms of 30-sec excerpts from the Bol-lywood and Hindustani genres respectively. For both plots the left half contains

Table 5. %Correct classification for different genres using fully automatic predominant F0 extraction; Bold indicates best achieved in each genre

Genre	I	II	III	IV	V	Total
Baseline	77.2	66.0	65.6	82.6	83.2	74.9
MFCCs(after source isolation)	76.9	72.3	70.0	78.9	83.0	76.2
C1	81.1	67.8	74.8	78.9	84.5	77.4
C2	72.0	75.9	77.1	67.5	65.8	71.7
C3	65.6	69.1	62.1	77.1	66.0	68.0
C1+C2	**82.9**	78.5	82.8	79.5	**85.0**	81.7
C1+C3	81.5	72.9	77.6	**83.9**	84.9	80.2
C1+C2+C3	82.1	**81.1**	**83.5**	83.0	84.7	**82.8**

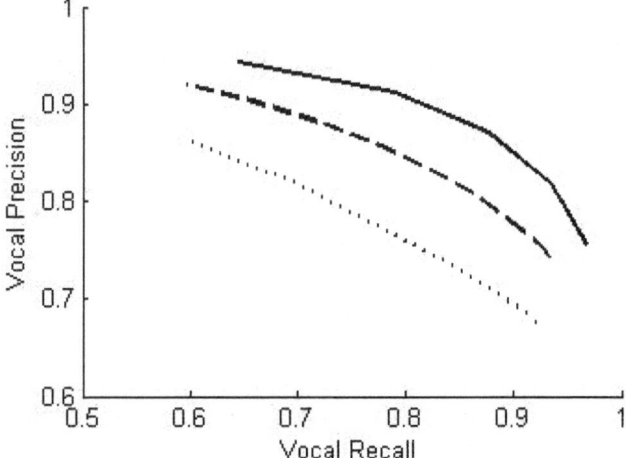

Fig. 1. Avg. Vocal Recall v/s Precision curves for different feature sets - baseline, C1 and C1+C2+C3 (using semi-automatic F0 extraction) across genres in the Leave 1 song out classification experiment

a dominant melodic instrument and right half contains vocals. In the Bollywood case the instrument is replete with large pitch modulations but the vocal part has mainly flatter note-pitches. However the instrumental timbre is largely invariant while the vocal part contains several phonemic transitions. In this case the timbral dynamic feature set (C2) is able to discriminate between the voice and instrument but the F0-harmonic dynamics feature set fails. The situation is reversed for the Hindustani excerpt since, although the instrumental part still displays timbral invariance, this is also exhibited by the vocal part, which consists of a long single utterance i.e. rapid pitch modulations on the held-out vowel /a/. C2 is ineffective in this case due to the absence of phonetic transitions. However the relative flatness of the instrument harmonics as compared with the vocal harmonics leads to good performance for C3.

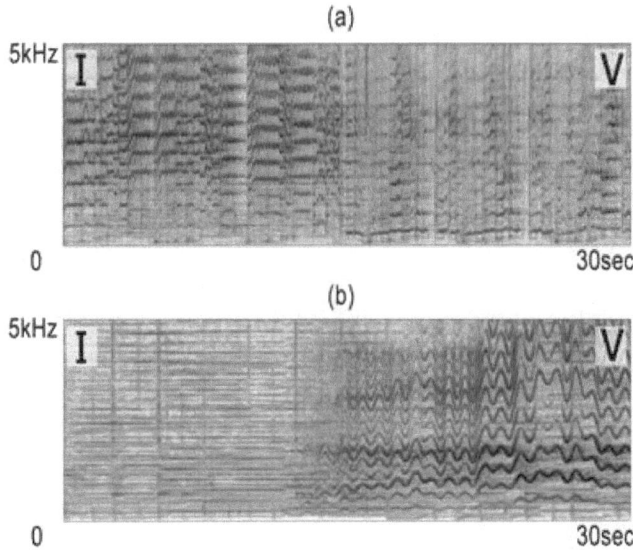

Fig. 2. Spectrograms of excerpts from (a) Bollywood and (b) Hindustani genres. For both excerpts the left section contains a dominant melodic instrument (I) and the right section contains vocals (V).

6 Conclusions

In this paper we have investigated the use of a combination of static and dynamic features for effective detection of lead vocal segments within polyphonic music in a cross-cultural context. Several of the features are novel and have been motivated by considering the distinctive characteristics of singing voice across genres. The introduction of an isolated dominant source spectral representation resulted in a significant increase in the performance of static timbral features in the polyphonic setting over a popularly used baseline feature set. The features representing timbral dynamics and F0-harmonic dynamics were found to provide complementary information for different underlying signal conditions related to singing styles and instrumentation specific to individual genres. While the overall combination of the static and dynamic features was found to result in the highest overall classification performance, individual genre accuracies clearly indicate the value of adapting feature sets to genre-specific acoustic characteristics. Thus commonly available metadata, such as genre, may be effectively utilized in the front-end of a MIR system.

References

1. Berenzweig, A., Ellis, D., Lawrence, S.: Using voice segments to improve artist classification of music. In: 22nd International Conference of Audio Engineering Society, Finland (2002)

2. Li, Y., Wang, D.: Separation of singing voice from music accompaniment for monoaural recordings. IEEE Trans. of Audio, Speech Lang. Proc. 15(4), 1475–1487 (2007)
3. Fujihara, H., Goto, M.: Three techniques for improving automatic synchronization between music and lyrics: Fricative detection, filler model and novel feature vectors for vocal activity detection. In: IEEE International Conference on Acoust., Speech, Signal Proc., Las Vegas (2008)
4. Lukashevich, H., Gruhne, M., Dittmar, C.: Effective singing voice detection in popular music using ARMA filtering. In: 10th International Conference on Digital Audio Effects (DAFx 2007), Bordeaux, France (2007)
5. Xiao, L., Zhou, J., Zhang, T.: Using DTW based unsupervised segmentation to improve the vocal part detection in pop music. In: IEEE International Conference on Multimedia and Expo, Hannover, Germany (2008)
6. Fujihara, et al.: F0 estimation method for singing voice in polyphonic audio signal based on statistical vocal model and viterbi search. In: IEEE International Conference on Acoust. Speech and Sig. Processing, Toulouse, France (2006)
7. Berenzweig, A., Ellis, D.: Locating singing voice segments within music signals. In: IEEE Workshop Applications of Sig. Process. to Audio and Acoust., New York (2001)
8. Maddage, N., Xu, C., Wang, Y.: A SVM-based classification approach to musical audio. In: International Conference on Music Information Retrieval, Baltimore (2003)
9. Ramona, M., Richard, G., David, B.: Vocal detection in music with support vector machines. In: IEEE International Conference on Acoust. Speech and Sig. Process. (2008)
10. Nwe, T., Li, H.: Exploring vibrato-motivated acoustic features for singer identification. IEEE Trans. Audio Speech Lang. Process. 15(2), 519–530 (2007)
11. Kim, Y., Whitman, B.: Singer identification in popular music recordings using voice coding features. In: Proc. 5th Intl. Conf. on Music Information Retrieval, Spain (2004)
12. Nwe, T., Li, H.: On fusion of timbre-motivated features for singing voice detection and singer identification. In: IEEE International Conference Acoust., Speech, Signal Proc., Las Vegas (2008)
13. Chou, W., Gu, L.: Robust singing detection in speech/music discriminator design. In: IEEE International Conference Acoust. Speech Sig. Process. (2001)
14. Tzanetakis, G.: Song-specific bootstrapping of singing voice structure. In: IEEE International Conference Multimedia and Expo, Taipei, Taiwan (2004)
15. Zhang, T.: System and method for automatic singer identification. In: IEEE International Conference Multimedia and Expo, Baltimore (2003)
16. Vallet, F., McKinney, M.: Perceptual constraints for automatic vocal detection in music recordings. In: Conference Interdisciplinary Musicology (2007)
17. Regnier, L., Peeters, G.: Singing voice detection in music tracks using direct voice vibrato detection. In: IEEE International Conference Acoust. Speech Sig. Process., Taipei, Taiwan (2009)
18. Lidy, T., et al.: On the Suitability of State-of-the-art Music Information Retrieval Methods for Analyzing, Categorizing and Accessing Non-Western and Ethnic Music Collections. In: Elsevier Signal Processing Special issue on Ethnic Music Audio Documents: From the Preservation to the Fruition (2009)
19. Mohammed, N., Squire, D.M.: Effectiveness of ICF features for collection-specific CBIR. In: Detyniecki, M., García-Serrano, A., Nürnberger, A., Stober, S. (eds.) AMR 2011. LNCS, vol. 7836, pp. 83–95. Springer, Heidelberg (2013)

20. Proutskova, P., Casey, M.: You call that singing? Ensemble classification for multi-cultural collections of music recordings. In: 10th International Conference on Music Information Retrieval, Kobe, Japan (2009)

21. Fuhrmann, F., Haro, M., Herrera, P.: Scalability, Generality and Temporal Aspects in Automatic Recognition of Predominant Musical Instruments in Polyphonic Music. In: 10th International Conference on Music Information Retrieval, Kobe, Japan (2009)

22. Fuhijara, H., Goto, M., Kitahara, T., Okuno, H.: A modeling of singing voice robust to accompaniment sounds and its application to singer identification and vocal-timbre-similarity-based music information retrieval. IEEE Trans. Audio, Speech, Lang. Process. 18(3), 638–648 (2010)

23. Rao, V., Rao, P.: Vocal melody extraction in the presence of pitched accompaniment in polyphonic music. IEEE Trans. Audio Speech and Lang. Process. 18(8), 2145–2154 (2010)

24. Pant, S., Rao, V., Rao, P.: A melody detection user interface for polyphonic music. In: National Conference Comm., Chennai, India (2010)

25. Rao, V., Gaddipati, P., Rao, P.: Signal-driven adaptation for singing voice processing in polyphony. IEEE Trans. Audio, Speech, Lang. Process. (2011) (accepted with minor mandatory revisions)

26. Serra, X.: Music sound modeling with sinusoids plus noise. In: Roads, C., Pope, S., Picialli, A., De Poli, G. (eds.) Musical Signal Processing, Swets and Zeitlinger (1997)

27. Rocamora, M., Herrera, P.: Comparing audio descriptors for singing voice detection in music audio files. In: Brazilian Symposium on Computer Music (2007)

28. Peeters, G.: A large set of audio features for sound description (similarity and classification) in the CUIDADO project. In: CUIDADO I.S.T. Project Report (2004)

29. Lagrange, M., Raspaud, M., Badeau, R., Richard, G.: Explicit modeling of temporal dynamics within musical signals for acoustic unit similarity. Pattern Recog. Letters 31(12), 1498–1506 (2010)

30. Burred, J., Robel, A., Sikora, T.: Dynamic spectral envelope modeling for timbre analysis of musical instrument sounds. IEEE Trans. Audio Speech Lang. Process. 18(3), 663–674 (2010)

31. Aucouturier, J.-J., Patchet, F.: The influence of polyphony on the dynamic modeling of musical timbre. Pattern Recog. Letters 28(5), 654–661 (2007)

32. Sundberg, J.: A rhapsody on perception. In: The Science of Singing Voice. Northern Illinois University Press (1987)

33. Shenoy, A., Wu, Y., Wang, Y.: Singing voice detection for karaoke application. In: Visual Comm. and Image Proc., Beijing, China (2005)

34. Rao, V., Rao, P.: Singing voice detection using predominant pitch. In: InterSpeech, Brighton, U.K. (2009)

35. Hall, M., et al.: The WEKA Data Mining Software: An Update. SIGKDD Explorations 11(1) (2009)

36. Bouman, C.: Cluster: An unsupervised algorithm for modeling Gaussian mixtures, http://www.ece.purdue.edu/~bouman

37. Kittler, J., Hatef, M., Duin, R., Matas, J.: On combining classifiers. IEEE Trans. Pattern Analysis and Machine Intelligence 20(3) (1998)

38. Markaki, M., Holzapfel, A., Stylianou, Y.: Singing voice detection using modulation frequency features. In: Workshop on Statistical and Perceptual Audition (2008)

Personalization in Multimodal Music Retrieval

Markus Schedl and Peter Knees

Department of Computational Perception
Johannes Kepler University
Linz, Austria
http://www.cp.jku.at

Abstract. This position paper provides an overview of current research endeavors and existing solutions in multimodal music retrieval, where the term "multimodal" relates to two aspects. The first one is taking into account the *music context* of a piece of music or an artist, while the second aspect tackled is that of the *user context*. The music context is introduced as all information important to the music, albeit not directly extractable from the audio signal (such as editorial or collaboratively assembled meta-data, lyrics in textual form, cultural background of an artists, or images of album covers). The user context, in contrast, is defined by various external factors that influence how a listener perceives music. It is therefore strongly related to user modeling and personalization, both facets of music information research that have not gained large attention by the MIR community so far. However, we are confident that adding personalization aspects to existing music retrieval systems (such as playlist generators, recommender systems, or visual browsers) is key to the future of MIR. In this vein, this contribution aims at defining the foundation for future research directions and applications related to multimodal music information systems.

1 Introduction and Motivation

Multimodal music processing and retrieval can be regarded as subfields of music information research, a discipline that has substantially gained importance during the last decade. The article at hand focuses on certain aspects of this field in that it will give an overview of the state-of-the-art in modeling and determining properties of music and listeners using features of different nature. In this introductory part, first a broad classification of such features is presented. Second, the three principle ways of *music retrieval* are introduced, each together with references to existing systems. Third, existing work on including personalization aspects in typical MIR tasks is reviewed.

The subsequent section points out various research endeavors and directions deemed to be important by the authors for the future of music information research, in particular, how to bring personalization and user adaptation to MIR. To this end, various data sources to describe the user context are introduced and discussed. Then, we present six steps towards the creation of a personalized multimodal music retrieval system.

M. Detyniecki et al. (Eds.): AMR 2011, LNCS 7836, pp. 58–71, 2013.

1.1   Categories of Features

Estimating perceived musical similarity is commonly achieved by describing aspects of the music entity (e.g., a song, a performer, or an album) or the listener via computational features, and employing a similarity measure. These features can be broadly categorized into three classes, according to the authors: *music content*, *music context*, and *user context*, cf. Figure 1.

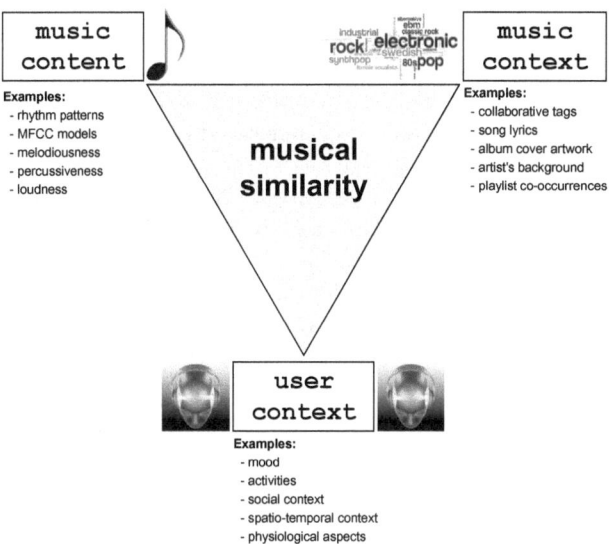

Fig. 1. Feature categories to describe music and listeners

Music Content. In content-based MIR, features extracted by applying signal processing techniques to audio signals are dominant. Such features are commonly denoted as *signal-based*, *audio-based*, or *content-based*. A good overview of common extraction techniques is presented in [7]. Music content-based features may be low-level representations that stem directly from the audio signal, for example zero-crossing rate [18], amplitude envelope [5], bandwidth and band energy ratio [37], or spectral centroid [67]. Alternatively, audio-based features may be derived or aggregated from low-level properties, and therefore represent aspects on a higher level of music understanding. Models of the human auditory system are frequently included in such derived features. High-level features usually aim at capturing either *timbral aspects* of music, which are commonly modeled via *MFCCs* [2], or *rhythmic aspects*, for example described via beat histograms [75] or fluctuation patterns [63,56]. Recent work addresses more specific high-level concepts, such as melodiousness and aggressiveness [57,52].

Music Context. The *music context* can be described as all information relevant to the music, albeit not directly extractable from the audio signal. For example, the meaning of a song's lyrics [29,26], the political background of the musician, or the geographic origin of an artist [19,66,65] are likely to have a large impact on the music, but are not manifested in the signal.

An overview of the state-of-the-art in *music context*-based feature extraction (and similarity estimation) can be found in [61]. The majority of the approaches covering the music context are strongly related to *Web content mining* [38] as the Web provides contextual information on music artists in abundance. For example, in [21] the authors construct term profiles created from *artist-related Web pages* to derive music similarity information. *RSS feeds* are extracted and analyzed in [8]. Alternative sources to mine music context-related data include *playlists* (e.g., radio stations and mix tapes, i.e., user-generated playlists) [3,6,48] and *Peer-to-Peer networks* [70,39,11,77]. In these cases, *co-occurrence analysis* is commonly employed to derive similarity on the artist- or track-level. Co-occurrences of artist names on Web pages are also used to infer artist similarity information [62] and for artist-to-genre classification [64]. *Song lyrics* as a source of music context-related information are analyzed, for example, in [40] to derive similarity information, in [33] for mood classification, and in [42] for genre classification. Another source for the music context is *collaborative tags*, mined for example from *last.fm* [32] in [12,36] or gathered via *tagging games* [41,74,34].

User Context. Existing work on incorporating user context aspects into MIR systems is relatively sparse. A preliminary study on users' acceptance of context logging in the context of music applications was conducted by Nürnberger and Stober [72]. The authors found significant differences in the participants' willingness to reveal different kinds of personal data on various scopes. Most participants indicated to eagerly share music metadata, information about ambient light and noise, mouse and keyboard logs, and their status in instant messaging applications. When it comes to used applications, facial expressions, bio signals, and GPS positions, however, a majority of users are reluctant to share their data. As for country-dependent differences, US-Americans were found to have on overall much lesser reservations to share personal data than Germans and Austrians. One has to note, however, that the study is biased towards Germans (accounting for 70% of the 305 participants).

In [59] Pohle et al. present preliminary steps towards a simple personalized music retrieval system. Based on a clustering of community-based tags extracted from *last.fm*, a small number of musical concepts are derived using *Non-Negative Matrix Factorization* (NMF) [35,78]. Each music artist or band is then described by a "concept vector". A user interface allows for adjusting the weights of the individual concepts, based on which artists that match the resulting distribution of the concepts best are recommended to the user. Zhang et al. propose in [80] a very similar kind of personalization strategy via user-adjusted weights.

Knees and Widmer present in [27] an approach that incorporates *relevance feedback* [60] into a text-based music search engine [23] to adapt the retrieval process to user preferences.

Even though no detailed information on their approach is publicly available, *last.fm* [32] builds user models based on its users' listening habits, which are mined via the "AudioScrobbler" interface. Based on this data, *last.fm* offers personalized music recommendations and playlist generation, however, without letting the user control (or even know) which factors are taken into account.

1.2   Categorizing Music Retrieval Systems

According to [76], music information retrieval systems to access music collections can be broadly categorized with respect to the employed *query formulation* method into *direct querying*, *query by example*, and *browsing* systems.

Direct querying systems take as input an excerpt of the feature representation to search for a piece of music. To give an example, *Themefinder* [73] and *musipedia* [44] support queries for sequences of exact pitch and of intervals, as well as for gross contour, using only up/down/repeat to describe the sequence of pitch changes.

A popular instance of a *query by example* system is *Shazam* [71], where the user records part of a music piece via his or her mobile phone, which is then analyzed on a server, identified, and meta-data such as artist or track name is sent back to the user's mobile phone. Another category of query by example retrieval applications is *query by humming* systems [13,30,49], where the search input consists of a user's recorded voice.

User interfaces that address the modality of *browsing* music collections exist in a considerable quantity. A fairly popular visualization and browsing technique employs the "Islands of Music" [53,50] metaphor, which uses *Self-Organizing Maps* (SOM) [28], i.e., a non-linear, topology-preserving transform of a high-dimensional feature space to usually two dimensions. There exist also various extensions to the basic "Island of Music" approach. For example, [51] present "Aligned SOMs" that allow a smooth shift between SOMs generated on features representing diametric aspects of (music) properties. A mobile version of the "Islands of Music" is presented in [46]. This version also features a simple playlist generation method. In [25,24] a three-dimensional extension is proposed to explore music collections in a fun way by further incorporating additional material mined from the Web. In addition, this three-dimensional version features intuitive playlist generation. It further makes use of an approach that is called "Music Description Map" [22] to calculate a mapping from music-related terms gathered from the Web to a SOM grid.

A browsing approach that offers an input wheel to sift through a cyclic playlist generated based on audio similarity is presented in [58]. A variant enriched with Web data and implemented on an "iPod touch" can be found in [68].

"The World of Music" [14] represents an appealing music artist visualizer and browser, which calculates an embedding of high-dimensional data into the visualization space by employing *Semidefinite Programming* (SDP) [15]. *Multidimensional Scaling* (MDS) [31,10] to visualize similar artist relations and browse in music collections is employed in [69]. Seyerlehner uses k-nearest neighbor graphs to reduce the computational complexity involved when dealing with medium- to large-sized collections and calculating a projection from the high-dimensional feature space to the two-dimensional visualization plane. Other interesting user interfaces for music collections include "MusicSun" [55], "MusicRainbow" [54], and "Musicream" [17].

1.3   Personalization Approaches

Aspects of the user context (cf. Section 1.1) are seldom taken into account when it comes to accessing music collections. One of the few commercial examples where the user context is considered in music search is the *collaborative filtering* [4] approach employed in *amazon.com*'s music Web store [1]. However, no details of the exact approach are publicly available.

In [9] Chai and Barry present some general considerations on modeling the user in a music retrieval systems and suggest an XML-based user modeling language for this purpose.

[47] presents a variant of the *Self-Organizing Map* (SOM) [28] that is based on a model that adapts to *user feedback*. To this end, the user can move data items on the SOM. This information is fed back into the SOM's codebook, and the mapping is adapted accordingly. [79] presents a *collaborative personalized search model* that alleviates the problems of *data sparseness* and *cold-start for new users* by combining information on different levels (individuals, interest groups, and global).

[80,81] present *CompositeMap*, a model that takes into account similarity aspects derived from music content as well as from social factors. The authors propose a multimodal music similarity measure and show its applicability to the task of music retrieval. They also allow a simple kind of personalization of this model by letting the user weight the individual music dimensions on which similarity is estimated. However, they neither take the user context into consideration, nor do they try to learn a user's preferences.

In [43] a multimodal music similarity model at the artist-level is proposed. To this end, the authors calculate a *partial order embedding* using *kernel functions*. Music context- and content-based features are combined by this means. However, this model does not incorporate any personalization strategies.

2   User Modeling and Personalization in Music Retrieval

User profiling is without doubt key to enable personalized music services of the future. In the past, typical MIR applications, such as automated playlist generators or music browsers, employed approaches based on similarity measures computed on features derived from some representation of the music or artist, for example, acoustic properties extracted from the audio signal [57] or term profiles calculated from music-related texts [21]. However, such approaches are known to be limited in their performance by some upper bound [2]. Furthermore, such approaches fall short of addressing the subjective component of music perception. What is it, for example, that makes you like a particular song when you are relaxing on Sunday morning? Do you prefer listening to happy or melancholic music when you are in a depressive mood? Which song do you relate to the first date with your beloved? The answer to these questions is most likely to be highly dependent on subjective factors. The sole use of music content- and music context-features described above is therefore insufficient to answer them.

That is where user modeling, personalization, and preference learning come into play. Models that combine different representation levels (e.g., low-level acoustic features and semantically meaningful tags) on different levels of data aggregation (e.g., segments within a piece of music, track-, artist-, or genre-level) and relate them to user profiles are crucial to describe user's preferences. The user model itself can also incorporate data on different levels of user representation. For example, [79] proposes a user model that comprises an *individual model*, a *interest group model*, and a *global user model*. We suggest adding a forth model, namely a *cultural user model*, that reflects the cultural area of the user. This cultural context can be given by an agglomeration, a whole country, or by a region that form a more or less homogeneous cultural entity.

2.1   Data Sources

There exists a wide variety of data sources for user context data, ranging from general location data (obtained by GPS or WiFi access points, for example) to highly personal aspects, such as blood pressure or intimate messages revealed by a user of a chatting software. Therefore, privacy issues play an important role for the acceptance of personalization techniques.

[16] provide a possible categorization of user context data. According to the authors, such data can be classified into the following five groups:

1. Environment Context
2. Personal Context
3. Task Context
4. Social Context
5. Spatio-temporal Context

The environmental context is defined as the entities that surround the user, for example, people, things, temperature, noise, humidity, and light. The personal context is further split into two sub-groups, namely, the *physiological context* and the *mental context*, where the former refers to attributes such as weight, blood pressure, pulse, or eye color, whereas the latter describes the user's psychic aspects, for example, stress level, mood, or expertise. The current activities pursued by the user is described by the *task context*. This context thus comprises actions, activities, and events the user is taking part of. Taking into account today's mobile phones with multi-tasking capabilities, we suggest to extend this definition to include aspects of direct user input, running applications, and information on which application currently has the focus. Further taking the different messenger and microblogging services of the Web 2.0 era into consideration, we propose including them into the category of task context. These services, however, may also be a valuable source for a user's *social context*, which gives information about relatives, friends, enemies, or collaborators. Finally, the *spatio-temporal context* reveals information about a user's location, place, direction, speed, and time.

The recent emergence of "always on" devices equipped not only with a permanent Web connection, but also with various built-in sensors, has remarkably facilitated the logging of user context data from a technical perspective. Integrated GPS modules, accelerometers, light and noise sensors as well as interfaces to almost every Web 2.0 service makes user context logging easier than ever before, by providing data for all context categories described above.

2.2   Towards Personalized Music Services

We believe the following steps to be crucial to establish a foundation for personalized music retrieval.

1. Investigate the suitability and acceptance of different data sources to create user profiles.
2. Develop methods to mine the most promising data sources.
3. Create a model that reflects different aspects of the user context.
4. Investigate directions to integrate different similarity measures (content-based, music context-based, and user context-based).
5. Develop and thoroughly evaluate integrated models for the three kinds of music similarity.
6. Build a user-adaptive system for music retrieval, taking into account user-related factors.

The first step is to investigate the user's readiness to disclose various kinds of user-specific information, which will contribute to creating a user profile. Such a model is indispensable for personalized music recommendation that reflects various aspects of the music, the listener, and his or her environment. For example, aspects such as current position and direction of movement (e.g., is the user at home, doing sports, driving a car, in a train), weather conditions, times of the day, activities he or she is pursuing while listening to music, his or her current mood and emotion, demographic and socio-economic information about the user, information on the used music playback device (e.g., size, power, storage capacity, battery status), and information on music files (e.g., audio features, cultural meta-data extracted from the Web, editorial meta-data published by record companies, personal meta-data like playcounts or user tags) contribute to how a user judges the similarity between two artists or songs. We assume that the user's willingness to disclose partly private and sensible information, such as geographic location, listening habits, Web browser histories and bookmarks, or content of shared folders in Peer-to-Peer networks, is strongly influenced by the benefits he or she can gain thereby (as one can easily see when looking at the overwhelming success of social networks). However, this willingness needs to be thoroughly evaluated, for example, by means of questionnaires and Web surveys.

Based on the results of the first step, it is possible to identify the most promising data sources, which a wide range of users are untroubled to share. Hence, the objective of the next step is to develop various data extractor components to gather user information, ranging from simple ones like date, time, and weekday monitoring, or recording user location and mouse clicking rates to complex ones such as bio-feedback measurements or user postings on social networks. For most data sources, employing post-processing to the gathered data will be required. To give an example from the Web mining domain, a study conducted in [20] revealed that about 50% of all user comments on *MySpace* [45] pages of popular music artists consists solely of spam, and 75% of the non-spam content failed linguistic parsing, meaning that 75% consists of broken sentences.

Step three subsequently aims at investigating which kinds of user context features relate to a listener's music taste, and at designing a user model that reflects and aggregates these user-specific factors. To this end, it is necessary to apply and refine machine learning techniques to learn user preferences, i.e., a mapping between individual, user-specific factors and the user's appeal to certain music categories, styles, or individual artists or tracks. In this step, various models of different scope and complexity need to be evaluated: for example, one model that takes only directly user-related data into account, another one similar to [79] that represents an integrated model comprising of an *individual user model*, a *group model* (cultural / peer group), and a *global model*.

Existing multifaceted models for music similarity measurement, such as [43,80], seem to lack real personalization functionality beyond simple user-adjustable weights for certain feature dimensions. Therefore, looking into different ways of building an aggregate model of music similarity based on the three broad categories of sources (music content, music context, and user context) is the key part of step four. Besides the problem of dealing with the inhomogeneous nature of the data sources, another important issue to address is the dimensionality problem since some data sources (term profiles in the case of user tags or Web page content, for example) are very high-dimensional, and therefore require the application and evaluation of dimensionality reduction techniques.

Following different strategies to develop such a comprehensive, multifaceted model will result in various model prototypes. In the next step, these prototypes have to undergo a comprehensive evaluation, including user studies and Web surveys. The best performing models are then determined for various usage scenarios, e.g., recommender systems, playlist generation, or retrieval systems supporting very specific, cross-data source queries such as "give me music for listening to on my mobile device when I am driving my car (user context), that further has a strong harmonic component (music content) and sad lyrics (music context)".

The final step comprises creating, evaluating, and refining various prototypical music retrieval systems that adapt to the user's current listening preferences, which are derived from the user context. The systems will make use of the aggregate models of music similarity elaborated in step five. They may include automatic personalized playlist generation systems, personalized recommender systems, or adaptive user interfaces to music collections. In this step, evaluating the ergonomic as well as the qualitative aspects of the retrieval systems is necessary.

3   User-Awareness and Personalization Are the Future of MIR

From the analysis and considerations presented so far, the authors' perspective of future research directions and music services can be summarized as follows.

Personalization aspects have to be taken into account when elaborating music retrieval systems. In this context, it is important to note the highly subjective, cognitive component in the understanding of music and judging its personal appeal. Therefore, designing user-aware music applications requires intelligent machine learning techniques, in particular, preference learning approaches that relate the user context to concise, situation-dependent music preferences.

User models that encompass different social scopes are needed. They may aggregate an individual model, an interest group model, a cultural model, and a global model.

Multifaceted similarity measures that combine different feature categories (music content, music context, and user context) are required. The corresponding representation models should then not only allow to derive similarity between music via content-related aspects, such as beat strength or instruments playing, or via music context-related properties, such as the geographic origin of the performer or a song's lyrics, but also to describe users and user groups in order to compute a listener-based similarity score. Such user-centric features enable the application of collaborative filtering techniques and eventually the elaboration of personalized music recommender systems.

Evaluation of user-adaptive systems is of vital importance. As such systems are by definition tailored to individual users, this is certainly not an easy task and far beyond the genre-classification-experiments commonly employed when assessing music similarity measures.

Nevertheless, we are sure that future research directions in MIR should be centered around intelligently combining various complementary music and user representations as this will pave the way for exciting novel music applications that keep on playing music according to the user's taste without requiring any explicit user interaction.

Acknowledgments. This research is supported by the Austrian Science Funds (FWF): P22856-N23.

References

1. http://www.amazon.com/music (access January 2010)
2. Aucouturier, J.-J., Pachet, F.: Improving Timbre Similarity: How High is the Sky? Journal of Negative Results in Speech and Audio Sciences 1(1) (2004)
3. Baccigalupo, C., Plaza, E., Donaldson, J.: Uncovering Affinity of Artists to Multiple Genres from Social Behaviour Data. In: Proceedings of the 9th International Conference on Music Information Retrieval (ISMIR 2008), Philadelphia, PA, USA, September 14-18 (2008)
4. Breese, J.S., Heckerman, D., Kadie, C.: Empirical Analysis of Predictive Algorithms for Collaborative Filtering. In: Proceedings of the 14th Conference on Uncertainty in Artificial Intelligence (UAI 1998), pp. 43–52. Morgan Kaufmann, San Francisco (1998)
5. Burred, J.J., Lerch, A.: A Hierarchical Approach to Automatic Musical Genre Classification. In: Proceedings of the 6th International Conference on Digital Audio Effects (DAFx 2003), London, UK, September 8-11 (2003)
6. Cano, P., Koppenberger, M.: The Emergence of Complex Network Patterns in Music Artist Networks. In: Proceedings of the 5th International Symposium on Music Information Retrieval (ISMIR 2004), Barcelona, Spain, October 10-14, pp. 466–469 (2004)
7. Casey, M.A., Veltkamp, R., Goto, M., Leman, M., Rhodes, C., Slaney, M.: Content-Based Music Information Retrieval: Current Directions and Future Challenges. Proceedings of the IEEE 96, 668–696 (2008)
8. Celma, O., Ramírez, M., Herrera, P.: Foafing the Music: A Music Recommendation System Based on RSS Feeds and User Preferences. In: Proceedings of the 6th International Conference on Music Information Retrieval (ISMIR 2005), London, UK, September 11-15 (2005)
9. Chai, W., Vercoe, B.: Using user models in music information retrieval systems. In: Proceedings of the International Symposium on Music Information Retrieval (ISMIR 2000), Plymouth, MA, USA (2000)
10. Cox, T.F., Cox, M.A.A.: Multidimensional Scaling. Chapman & Hall (1994)
11. Ellis, D.P., Whitman, B., Berenzweig, A., Lawrence, S.: The Quest For Ground Truth in Musical Artist Similarity. In: Proceedings of 3rd International Conference on Music Information Retrieval (ISMIR 2002), Paris, France, October 13-17 (2002)
12. Geleijnse, G., Schedl, M., Knees, P.: The Quest for Ground Truth in Musical Artist Tagging in the Social Web Era. In: Proceedings of the 8th International Conference on Music Information Retrieval (ISMIR 2007), Vienna, Austria, September 23-27 (2007)
13. Ghias, A., Logan, J., Chamberlin, D., Smith, B.C.: Query by Humming: Musical Information Retrieval in an Audio Database. In: Proceedings of the 3rd Association for Computing Machinery (ACM) International Conference on Multimedia, San Fancisco, CA, USA, pp. 231–236 (1995)
14. Gleich, D., Rasmussen, M., Lang, K., Zhukov, L.: The World of Music: SDP Layout of High Dimensional Data. In: Proceedings of the IEEE Symposium on Information Visualization 2005 (2005)
15. Goemans, M.X., Williamson, D.P.: Improved Approximation Algorithms for Maximum Cut and Satisfyability Problems Using Semidefinite Programming. Journal of the Association for Computing Machinery 42(6), 1115–1145 (1995)
16. Göker, A., Myrhaug, H.I.: User Context and Personalisation. In: Proceedings of the 6th European Conference on Case Based Reasoning (ECCBR 2002): Workshop on Case Based Reasoning and Personalization, Aberdeen, Scotland (September 2002)

17. Goto, M., Goto, T.: Musicream: New Music Playback Interface for Streaming, Sticking, Sorting, and Recalling Musical Pieces. In: Proceedings of the 6th International Conference on Music Information Retrieval (ISMIR 2005), London, UK, September 11-15 (2005)
18. Gouyon, F., Pachet, F., Delerue, O.: On the Use of Zero-Crossing Rate for an Application of Classification of Percussive Sounds. In: Proceedings of the COST-G6 Conference on Digital Audio Effects (DAFx 2000), Verona, Italy, December 7-9 (2000)
19. Govaerts, S., Duval, E.: A Web-based Approach to Determine the Origin of an Artist. In: Proceedings of the 10th International Society for Music Information Retrieval Conference (ISMIR 2009), Kobe, Japan (October 2009)
20. Grace, J., Gruhl, D., Haas, K., Nagarajan, M., Robson, C., Sahoo, N.: Artist Ranking Through Analysis of On-line Community Comments. In: Proceedings of the 17th ACM International World Wide Web Conference (WWW 2008), Bejing, China, April 21-25 (2008)
21. Knees, P., Pampalk, E., Widmer, G.: Artist Classification with Web-based Data. In: Proceedings of the 5th International Symposium on Music Information Retrieval (ISMIR 2004), Barcelona, Spain, October 10-14, pp. 517–524 (2004)
22. Knees, P., Pohle, T., Schedl, M., Widmer, G.: Automatically Describing Music on a Map. In: Proceedings of 1st Workshop on Learning the Semantics of Audio Signals (LSAS 2006), Athens, Greece, December 6-8 (2006)
23. Knees, P., Pohle, T., Schedl, M., Widmer, G.: A Music Search Engine Built upon Audio-based and Web-based Similarity Measures. In: Proceedings of the 30th Annual International ACM SIGIR Conference on Research and Development in Information Retrieval (SIGIR 2007), Amsterdam, the Netherlands, July 23-27 (2007)
24. Knees, P., Schedl, M., Pohle, T., Widmer, G.: An Innovative Three-Dimensional User Interface for Exploring Music Collections Enriched with Meta-Information from the Web. In: Proceedings of the 14th ACM International Conference on Multimedia (MM 2006), Santa Barbara, CA, USA, October 23-27 (2006)
25. Knees, P., Schedl, M., Pohle, T., Widmer, G.: Exploring Music Collections in Virtual Landscapes. IEEE MultiMedia 14(3), 46–54 (2007)
26. Knees, P., Schedl, M., Widmer, G.: Multiple Lyrics Alignment: Automatic Retrieval of Song Lyrics. In: Proceedings of 6th International Conference on Music Information Retrieval (ISMIR 2005), London, UK, September 11-15, pp. 564–569 (2005)
27. Knees, P., Widmer, G.: Searching for Music Using Natural Language Queries and Relevance Feedback. In: Boujemaa, N., Detyniecki, M., Nürnberger, A. (eds.) AMR 2007. LNCS, vol. 4918, pp. 109–121. Springer, Heidelberg (2008)
28. Kohonen, T.: Self-Organizing Maps, 3rd edn. Springer Series in Information Sciences, vol. 30. Springer, Berlin (2001)
29. Korst, J., Geleijnse, G.: Efficient lyrics retrieval and alignment. In: Verhaegh, W., Aarts, E., Ten Kate, W., Korst, J., Pauws, S. (eds.) Proceedings of the 3rd Philips Symposium on Intelligent Algorithms (SOIA 2006), Eindhoven, the Netherlands, December 6-7, pp. 205–218 (2006)
30. Kosugi, N., Nishihara, Y., Sakata, T., Yamamuro, M., Kushima, K.: A Practical Query-by-Humming System for a Large Music Database. In: Proceedings of the 8th ACM International Conference on Multimedia, Los Angeles, CA, USA, pp. 333–342 (2000)
31. Kruskal, J.B., Wish, M.: Multidimensional Scaling. Paper Series on Quantitative Applications in the Social Sciences. Sage Publications, Newbury Park (1978)

32. http://last.fm (access January 2010)
33. Laurier, C., Grivolla, J., Herrera, P.: Multimodal Music Mood Classification using Audio and Lyrics. In: Proceedings of the International Conference on Machine Learning and Applications, San Diego, CA, USA (2008)
34. Law, E., von Ahn, L., Dannenberg, R., Crawford, M.: Tagatune: A Game for Music and Sound Annotation. In: Proceedings of the 8th International Conference on Music Information Retrieval (ISMIR 2007), Vienna, Austria (September 2007)
35. Lee, D.D., Seung, H.S.: Learning the Parts of Objects by Non-negative Matrix Factorization. Nature 401(6755), 788–791 (1999)
36. Levy, M., Sandler, M.: A semantic space for music derived from social tags. In: Proceedings of the 8th International Conference on Music Information Retrieval (ISMIR 2007), Vienna, Austria (September 2007)
37. Li, D., Sethi, I.K., Dimitrova, N., McGee, T.: Classification of General Audio Data for Content-based Retrieval. Pattern Recognition Letters 22(5), 533–544 (2001)
38. Liu, B.: Web Data Mining – Exploring Hyperlinks, Contents and Usage Data. Springer, Heidelberg (2007)
39. Logan, B., Ellis, D.P., Berenzweig, A.: Toward Evaluation Techniques for Music Similarity. In: Proceedings of the 26th Annual International ACM SIGIR Conference on Research and Development in Information Retrieval (SIGIR 2003): Workshop on the Evaluation of Music Information Retrieval Systems, Toronto, Canada, July-August. ACM Press (2003)
40. Logan, B., Kositsky, A., Moreno, P.: Semantic Analysis of Song Lyrics. In: Proceedings of the IEEE International Conference on Multimedia and Expo (ICME 2004), Taipei, Taiwan, June 27-30 (2004)
41. Mandel, M.I., Ellis, D.P.: A Web-based Game for Collecting Music Metadata. In: Proceedings of the 8th International Conference on Music Information Retrieval (ISMIR 2007), Vienna, Austria (September 2007)
42. Mayer, R., Neumayer, R., Rauber, A.: Rhyme and Style Features for Musical Genre Classification by Song Lyrics. In: Proceedings of the 9th International Conference on Music Information Retrieval (ISMIR 2008) (2008)
43. McFee, B., Lanckriet, G.: Heterogeneous Embedding for Subjective Artist Similarity. In: Proceedings of the 10th International Society for Music Information Retrieval Conference (ISMIR 2009), Kobe, Japan (October 2009)
44. http://www.musipedia.org (access February 2010)
45. http://www.myspace.com (access November 2009)
46. Neumayer, R., Dittenbach, M., Rauber, A.: PlaySOM and PocketSOMPlayer, Alternative Interfaces to Large Music Collections. In: Proceedings of the 6th International Conference on Music Information Retrieval (ISMIR 2005), London, UK, September 11-15 (2005)
47. Nürnberger, A., Detyniecki, M.: Weighted Self-Organizing Maps: Incorporating User Feedback. In: Kaynak, O., Alpaydın, E., Oja, E., Xu, L. (eds.) ICANN 2003 and ICONIP 2003. LNCS, vol. 2714, pp. 883–890. Springer, Heidelberg (2003)
48. Pachet, F., Westerman, G., Laigre, D.: Musical Data Mining for Electronic Music Distribution. In: Proceedings of the 1st International Conference on Web Delivering of Music (WEDELMUSIC 2001), Florence, Italy, November 23-24 (2001)
49. Padro, B.: Finding Structure in Audio for Music Information Retrieval. IEEE Signal Processing Magazine 23(3), 126–132 (2006)
50. Pampalk, E.: Islands of Music: Analysis, Organization, and Visualization of Music Archives. Master's thesis, Vienna University of Technology, Vienna, Austria (2001), http://www.oefai.at/~elias/music/thesis.html

51. Pampalk, E.: Aligned Self-Organizing Maps. In: Proceedings of the Workshop on Self-Organizing Maps (WSOM 2003), Kitakyushu, Japan, September 11-14, pp. 185–190. Kyushu Institute of Technology (2003)

52. Pampalk, E.: Computational Models of Music Similarity and their Application to Music Information Retrieval. PhD thesis, Vienna University of Technology (March 2006)

53. Pampalk, E., Dixon, S., Widmer, G.: Exploring Music Collections by Browsing Different Views. Computer Music Journal 28(3) (2004)

54. Pampalk, E., Goto, M.: MusicRainbow: A New User Interface to Discover Artists Using Audio-based Similarity and Web-based Labeling. In: Proceedings of the 7th International Conference on Music Information Retrieval (ISMIR 2006), Victoria, Canada, October 8-12 (2006)

55. Pampalk, E., Goto, M.: MusicSun: A New Approach to Artist Recommendation. In: Proceedings of the 8th International Conference on Music Information Retrieval (ISMIR 2007), Vienna, Austria, September 23-27 (2007)

56. Pampalk, E., Rauber, A., Merkl, D.: Content-based Organization and Visualization of Music Archives. In: Proceedings of the 10th ACM International Conference on Multimedia (MM 2002), Juan les Pins, France, December 1-6, pp. 570–579 (2002)

57. Pohle, T.: Automatic Characterization of Music for Intuitive Retrieval. PhD thesis, Johannes Kepler University Linz, Linz, Austria (2009)

58. Pohle, T., Knees, P., Schedl, M., Pampalk, E., Widmer, G.: "Reinventing the Wheel": A Novel Approach to Music Player Interfaces. IEEE Transactions on Multimedia 9, 567–575 (2007)

59. Pohle, T., Knees, P., Schedl, M., Widmer, G.: Building an Interactive Next-Generation Artist Recommender Based on Automatically Derived High-Level Concepts. In: Proceedings of the 5th International Workshop on Content-Based Multimedia Indexing (CBMI 2007), Bordeaux, France (2007)

60. Rocchio, J.J.: Relevance Feedback in Information Retrieval. In: Salton, G. (ed.) The SMART Retrieval System - Experiments in Automatic Document Processing, pp. 313–323. Prentice-Hall, Englewood Cliffs (1971)

61. Schedl, M., Knees, P.: Context-based Music Similarity Estimation. In: Proceedings of the 3rd International Workshop on Learning the Semantics of Audio Signals (LSAS 2009), Graz, Austria (2009)

62. Schedl, M., Knees, P., Widmer, G.: A Web-Based Approach to Assessing Artist Similarity using Co-Occurrences. In: Proceedings of the 4th International Workshop on Content-Based Multimedia Indexing (CBMI 2005), Riga, Latvia, June 21-23 (2005)

63. Schedl, M., Pampalk, E., Widmer, G.: Intelligent Structuring and Exploration of Digital Music Collections. e&i - Elektrotechnik und Informationstechnik 122(7-8), 232–237 (2005)

64. Schedl, M., Pohle, T., Knees, P., Widmer, G.: Assigning and Visualizing Music Genres by Web-based Co-Occurrence Analysis. In: Proceedings of the 7th International Conference on Music Information Retrieval (ISMIR 2006), Victoria, Canada, October 8-12 (2006)

65. Schedl, M., Schiketanz, C., Seyerlehner, K.: Country of Origin Determination via Web Mining Techniques. In: Proceedings of the IEEE International Conference on Multimedia and Expo (ICME 2010): 2nd International Workshop on Advances in Music Information Research (AdMIRe 2010), Singapore, July 19-23 (2010)

66. Schedl, M., Seyerlehner, K., Schnitzer, D., Widmer, G., Schiketanz, C.: Three Web-based Heuristics to Determine a Person's or Institution's Country of Origin. In: Proceedings of the 33rd Annual International ACM SIGIR Conference on Research and Development in Information Retrieval (SIGIR 2010), Geneva, Switzerland, July 19-23 (2010)
67. Scheirer, E., Slaney, M.: Construction and Evaluation of a Robust Multifeature Speech/Music Discriminator. In: Proceedings of the International Conference on Acoustics, Speech, and Signal Processing (ICASSP 1997), Munich, Germany, April 21–24, pp. 1331–1334 (1997)
68. Schnitzer, D., Pohle, T., Knees, P., Widmer, G.: One-Touch Access to Music on Mobile Devices. In: Proceedings of the 6th International Conference on Mobile and Ubiquitous Multimedia (MUM 2007), Oulu, Finland, December 12-14 (2007)
69. Seyerlehner, K.: Inhaltsbasierte Ähnlichkeitsmetriken zur Navigation in Musik-sammlungen. Master's thesis, Johannes Kepler Universität Linz, Linz, Austria (June 2006)
70. Shavitt, Y., Weinsberg, U.: Songs Clustering Using Peer-to-Peer Co-occurrences. In: Proceedings of the IEEE International Symposium on Multimedia (ISM 2009): International Workshop on Advances in Music Information Research (AdMIRe 2009), San Diego, CA, USA (December 16, 2009)
71. http://www.shazam.com (access February 2010)
72. Stober, S., Steinbrecher, M., Nürnberger, A.: A Survey on the Acceptance of Listening Context Logging for MIR Applications. In: Proceedings of 3rd Workshop on Learning the Semantics of Audio Signals (LSAS 2009), Graz, Austria (December 2009)
73. http://themefinder.org (access February 2010)
74. Turnbull, D., Liu, R., Barrington, L., Lanckriet, G.: A Game-based Approach for Collecting Semantic Annotations of Music. In: Proceedings of the 8th International Conference on Music Information Retrieval (ISMIR 2007), Vienna, Austria (September 2007)
75. Tzanetakis, G., Cook, P.: Musical Genre Classification of Audio Signals. IEEE Transactions on Speech and Audio Processing 10(5), 293–302 (2002)
76. Veltkamp, R.C.: Multimedia Retrieval Algorithmics. In: van Leeuwen, J., Italiano, G.F., van der Hoek, W., Meinel, C., Sack, H., Plášil, F. (eds.) SOFSEM 2007. LNCS, vol. 4362, pp. 138–154. Springer, Heidelberg (2007)
77. Whitman, B., Lawrence, S.: Inferring Descriptions and Similarity for Music from Community Metadata. In: Proceedings of the 2002 International Computer Music Conference (ICMC 2002), Göteborg, Sweden, September 16-21, pp. 591–598 (2002)
78. Xu, W., Liu, X., Gong, Y.: Document Clustering Based on Non-negative Matrix Factorization. In: Proceedings of the 26th Annual International ACM SIGIR Conference on Research and Development in Information Retrieval (SIGIR 2003), Toronto, Canada, July 28-August 1, pp. 267–273. ACM Press (2003)
79. Xue, G.-R., Han, J., Yu, Y., Yang, Q.: User Language Model for Collaborative Personalized Search. ACM Transactions on Information Systems 27(2) (February 2009)
80. Zhang, B., Shen, J., Xiang, Q., Wang, Y.: CompositeMap: A Novel Framework for Music Similarity Measure. In: Proceedings of the 32nd International ACM SIGIR Conference on Research and Development in Information Retrieval (SIGIR 2009), pp. 403–410. ACM, New York (2009)
81. Zhang, B., Xiang, Q., Wang, Y., Shen, J.: CompositeMap: A Novel Music Similarity Measure for Personalized Multimodal Music Search. In: MM 2009: Proceedings of the Seventeen ACM International Conference on Multimedia, pp. 973–974. ACM, New York (2009)

Classifying Images at Scene Level: Comparing Global and Local Descriptors

Christian Hentschel, Sebastian Gerke, and Eugene Mbanya

Fraunhofer Institute for Telecommunications, Heinrich Hertz Institute
{christian.hentschel,sebastian.gerke,eugene.mbanya}@hhi.fraunhofer.de

Abstract. In this paper we compare two state-of-the-art approaches for image classification. The first approach follows the Bag-of-Keypoints method for classifying images based on local image pattern frequency distribution. The second approach computes the gist of an image by computing global image statistics. Both approaches are explained in detail and their performance is compared using a subset of images taken from the ImageClef 2011 PhotoAnnotation task. The images were selected based on the assumption they could be better described using global features. Results show that while Bag-of-Keypoints-like classification performs better even for global concepts the classification accuracy of the global descriptor remains acceptable at a much smaller computational footprint.

1   Introduction

Visual data such as image and video represents the fastest growing data in the Internet today. Photo communities such as Flickr host more than 5 billion photos and community members upload more than 3000 images every minute[1]. Efficient retrieval methods are thus required to provide access to this vast amount of information, which would otherwise be useless. The sheer amount of data, however, renders manual annotation impossible. Automatic annotation of images has therefore become a keystone to semantic retrieval of visual data.

The concept behind all efforts is to automatically classify images into one or more predefined categories. This is preferably performed using only the pictorial information that is depicted in the images. Approaches in automatic classification employ techniques borrowed from Computer Vision to extract meaningful image features.

These features can be employed to separate one category of images from another by training a classifier using standard Machine Learning techniques.

Recently, the application of the *bag of keypoints* [2] or *visual codebook* model [6] has emerged as a de facto standard. The general idea behind this approach is motivated by an analogy to methods for text classification. A document is represented by a *bag-of-words*: a histogram counting the frequency of appearance of terms taken from a small representative codebook.

---

[1] Flickr, http://blog.flickr.net/en/2010/09/19/5000000000/

M. Detyniecki et al. (Eds.): AMR 2011, LNCS 7836, pp. 72–82, 2013.

In the visual domain, as there are no a priori *words* in an image, particular local image pattern are taken instead. Usually these pattern are described by histogram-of-gradient methods such as SIFT descriptors (Scale-Invariant-Feature-Transform, [7]). The location of these patterns is determined either by appropriate feature detection methods (e.g. Gaussian pyramid) or set to a fixed grid of sampling points. The advantage of the latter approach being that patterns are extracted even when the response of the selected feature detection algorithm is missing (e.g. due to missing or faint edges). The codebook is typically generated by vector quantizing the extracted set of descriptors into k representative clusters (or codewords). An image can then be described by a feature vector formed by assigning each local descriptor to its most similar codeword (or cluster center). Hence, the feature vector is a k-dimensional histogram, referred to as *bag of keypoints* [2] in analogy to the text classification approach[2].

Given a set of pre-classified training images, the feature vectors of these images can be used to train a classification model. Nowadays Support Vector Machines (SVM) typically serve as classification method. They usually learn a hyperplane that best separates the feature vectors of one image class (positive samples) from those of all images not belonging to this class (negative samples). Unseen images can be classified using the same kind of feature vectors and the trained SVM model. Figure 1 briefly outlines the general idea of this approach.

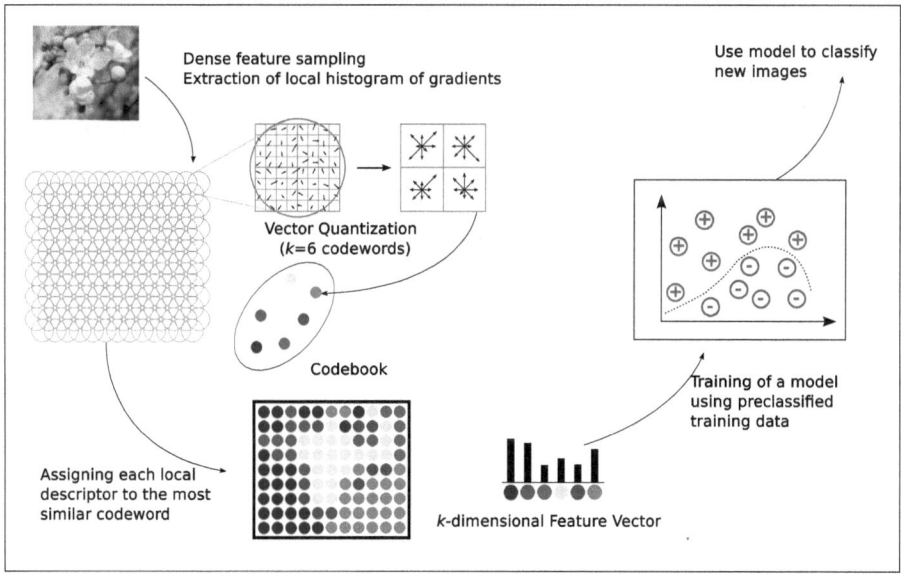

Fig. 1. Overview: Bag-of-words classification

---

[2] We will use the term *bag-of-words* throughout this document in order to emphasize that we compute SIFT descriptors at a fixed dense grid of sampling points.

The advantage of this concept is its simplicity, and its various invariances. The combination of SIFT (or similar) for local image description and the bag-of-words model makes it invariant to transformations, changes in lighting and rotation, occlusion, and intra-class variations [2]. By simply counting object features present in an image, missing or occluded object parts do not affect the classification accuracy in the same way as it would be the case for model-driven classification. Moreover, the approach is very generic. It can be extended to additional categories simply by training another model on manually labeled training data. The features do not have to be adapted to the classification task – once extracted they can be re-used.

The disadvantage, however, being that it totally neglects spatial image coherence. The layout of the image features is completely discarded as it cannot be captured by the histogram, which only counts the frequencies of the codebook vectors but not their spatial correlation. In this paper we therefore analyze to what extent this approach can be successful when it is applied to visual data that is better described "as a whole" as it is the case for images depicting sunsets or landscapes. It is questionable whether these scenic images can be ideally be described by means of image part frequency distributions.

The GIST descriptor has successfully applied to scene classification and recognition [9]. Scene is not considered as just an arrangement of objects but rather as an individual object itself with a unitary shape that is similar among scenes belonging to the same category. Computation of the descriptor involves accumulating image statistics over the entire scene rather than over local pattern.

In this paper we compare the object-oriented bag-of-words image representation with the scene-oriented GIST image representation. We intended to evaluate whether GIST in fact can outperform the bag-of-words approach when classifying images that could better be described at a global level. Evaluation was performed on the data provided by the ImageCLEF 2011 Photo Annotation Task.

The paper is structured as follows: Section 2 describes the process of bag-of-words extraction in more detail as well as an extension to so called spatial grids. Section 3 provides furthe details on the GIST descriptor. In section 4 we describe the process of learning image categories from both features using Support Vector Machines. Finally, section 5 gives niformation on the comparison of both descriptors as well as their combined overall performance and concludes this paper.

2   Features for Object Classification

As mentioned in the previous section, the bag-of-words model computes histograms from local image patterns such as SIFT. By default, SIFT operates on the luminance channel of an image. For color photos this would mean to drop two thirds of the available information. Experiments have shown that extending SIFT to color space largely improves classification accuracy. We therefore apply

SIFT to the opponent color space, a combination which has shown to provide good results in bag-of-words classification [12].

For that purpose, the image is converted to the opponent color space. The opponent color space is given by the following definition from the RGB color space:

$$\begin{pmatrix} O_1 \\ O_2 \\ O_3 \end{pmatrix} = \begin{pmatrix} \frac{R-G}{\sqrt{2}} \\ \frac{R+G-2B}{\sqrt{6}} \\ \frac{R+G+B}{\sqrt{3}} \end{pmatrix} \qquad (1)$$

Channel O_3 represents the intensity channel, while O_1 and O_2 represent the color components. Due to the subtraction in the first two channels, these are shift-invariant with respect to light intensity but not scale-invariant [12]. Since standard SIFT extracts histogram of gradients at local extrema of difference of Gaussians function applied in scale-space, regions with no or only few texture would not be used for descriptor computation. We therefore follow a dense sampling approach extracting a local descriptor every 6 pixel in a honeycomb grid. At each grid point SIFT descriptors are extracted for each of the three color channels resulting in three 128-dimensional descriptor. These are concatenated into one 384-dimension feature vector describing a local image pattern. We use the OpponentSIFT implementation provided by [13].

In order to compute a codebook we use k-means clustering for vector quantization. In this paper we used $k = 4000$ obtaining thus 4000 codewords. In order to reduce the computational effort for k-means clustering we randomly subsampled 800.000 features that have been actually used for codebook generation.

For each image we compute the histogram by assigning each extracted local descriptor to the most similar codeword using a most simple nearest neighbor classifier. Thus we obtain a 4000-dimensional feature vector per image.

2.1 Spatial Grids

An extension of the standard bag of keypoints approach was presented in [5]. As some concepts are more likely to be present in specific regions of an image (e.g. sky is more likely to be present in the upper image part) the authors suggest to split an image into fixed subregions. Different resolutions of subregions are aggregated into a so-called spatial grid. Histograms of codewords are computed for each of these subregions.

We follow this approach by defining two spatial partitions: one (1×3) consisting of 3 vertically stacked regions and one (2×2) consisting of the four image quadrants. Next to the 4000-bin histogram feature vector we already computed on the whole image, we hence obtain another 7: three for 1×3 partitioning and four for the 2×2 partitioning. The idea is illustrated in Fig. 2.

Finally, the histograms for a spatial partitioning are concatenated resulting in a total of 3 image feature vectors – one per grid resolution – at 4000, 12.000 and 16.000 dimensions. The feature vectors are then L_1 normalized, i.e. divided by the sum of the bin population.

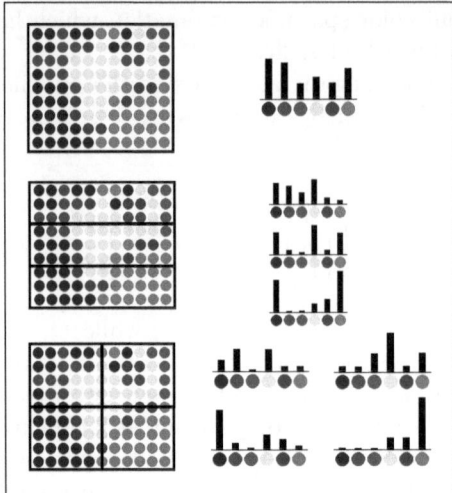

Fig. 2. Spatial Grids – an image is represented by histogram of codewords extracted at different subregions. Top: whole image, Middle: 1×3, Bottom: 2×2.

3   Features for Scene Classification

The spatial envelope model was initially introduced in [9] as a low-dimensional representation of a scene and successfully applied for k-nearest-neighbor scene classification. The idea was rather than applying segmentation or any kind of processing of individual objects or regions to represent the dominant spatial structure of a scene. The authors propose a set of perceptual dimensions (naturalness, openness, roughness, expansion, ruggedness) that represent the dominant spatial structure of a scene. The spatial envelop model was later termed GIST descriptor following Friedman's [4] definition of a scene gist, an abstract representation of the scene that spontaneously activates memory representations of scene categories (a city, a mountain, etc.), which is essentially what is captured by the spatial envelope model [8].

Extraction of GIST is performed by filtering the image by a bank of Gabor filters. The image is split into a 4×4 regular grid. Next, the Gabor filter responses are averaged over each block.

We use the implementation presented in [3] which takes as input a squared gray-level image of fixed size. All images are rescaled in a preliminary stage to 256×256 irrespective of their aspect ratio. We obtain a GIST feature vector of 512 dimensions per image (Gabor filter at 4 different scales, 8 filter orientations per scale), which is significantly less than the bag-of-words vectors at even the smallest grid resolution. Moreover, the GIST features can be computed much more efficiently as there is no step of codebook and histogram computation involved.

While bag-of-words is based on local gradient statistic, GIST employs frequency filtering

On the other hand, Gabor filter response at different scalings and orientations to some extent resembles the SIFT features as both focus on high frequency

4   Category Learning

Once the different feature vectors have been computed the problem of automatic image classification can be approached by machine learning techniques. Kernel-based Support Vector Machines (SVM) have been widely used in image classification scenarios (see e.g. [1,2,10,12,14]). In order to be able to predict the category of an unlabeled image the SVM classifier needs to be trained using labeled data. Based on the training data the SVM computes a decision boundary to distinguish categories.

We consider the classification task a one-against-all approach. We train one SVM per given image category to separate the images from this very category from all other given categories. Hence, the classifier is trained to solve a binary classification problem, i.e. whether or not an image belongs to a specific category or not.

This approach provides us with two advantages. First, new categories can easily be added by simply training a new classifier. Second, each image can be classified into multiple categories depending on the prediction confidence of each classifier available. We use SVM classifiers for both, bag-of-words features as well as GIST features.

4.1   Bag-of-Words Classification

We use a kernel function that is based on the χ^2 distance for training the Support Vector Machine on bag-of-words features. As mentioned the bag-of-words feature vectors are histograms of codeword distributions. The χ^2 distance has shown to provide good results for comparing distributions [14]. Given two histograms $H = (h_1, ..., h_m)$ and $H' = (h'_1, ..., h'_m)$ the χ^2 distance is defined as:

$$D(H, H') = \frac{1}{2} \sum_{i=1}^{m} \frac{(h_i - h'_i)^2}{|h_i| + |h'_i|} \tag{2}$$

To incorporate this metric into a Support Vector Machine we use a Gaussian kernel:

$$K(H, H') = exp(-\frac{1}{\gamma} D(H, H')) \tag{3}$$

The normalization parameter γ can be optimized using grid search and cross-validation. However, Zhang et al. [14] have shown, that setting this value to the average distance between all training image histograms gives comparable results and reduces the computational effort. Therefore, the only parameter we optimize in a cross-validation is the cost parameter C of the support vector classification.

As described in section 2.1, we yield three different histograms per image – one per spatial grid resolution. We compute a kernel matrix for each grid resolution.

4.2   GIST Classification

Similar to the bag-of-words features we train a SVM on GIST features. GIST feature vectors are compared using the standard Euclidean distance:

$$D(G, G') = \sum_{i=1}^{l} (g_i - g_i')^2 \Big.^{\frac{1}{2}} \tag{4}$$

where l is the dimensionality of the GIST descriptor (in our case $l = 512$).

We again use the Gaussian kernel for SVM classification, which in this case represents the standard radial basis function kernel (RBF). Similar to 4.1 we again only optimize the cost parameter of the SVM and set the kernel width γ to the average L2 distance between all training samples. While this has shown to be less optimal than in the bag-of-words case it still provides good results.

4.3   Classifier Fusion

Having computed 4 different kernels, naturally the next question is how to fuse them to yield better classification results. As we wanted to compare the performance of bag-of-words classification over GIST classification, we treated both features independently first, meaning that we fused the three bag-of-words kernels and compared their output to the GIST kernel.

There are various methods in literature that deal with the fusion of features at different levels. Considering kernel-based classification approaches, three different methods can be distinguished:

- *early fusion*, which refers to the concatenation of different feature vectors into a single vector
- *intermediate fusion*, which refers to the fusion at kernel level, e.g. creating the weighted sum
- *late fusion*, which refers to the fusion of the classifier outputs, e.g. the category confidences provided for a sample.

Early fusion typically requires very similar features that can be classified using the same distance measure and the same classification approach. As the bag-of-words features and the GIST features of our approach are compared using χ^2 and $L2$ norm respectively we could not apply early fusion techniques. Moreover we intended to fuse both features at kernel level.

Late fusion usually means to unite individual classifier outputs, i.e. concept probabilities into a common representation, typically similar to early fusion again by vector concatenation. These vectors are then used as input for a supervised learner, e.g. a support vector machine. A strong disadvantage of this concept is that it requires another training step with labeled training data which should be different from the dataset used to train the individual feature classifiers to prevent over-fitting. An advantage of this approach, however, is that it can exploit nonlinear relationships between individual classifier. Late fusion can also be performed unsupervised by combining the classifier outputs by arithmetic operations such as taking the maximum, the minimum, the average, or the geometric mean of individual classifier probabilities. This avoids the need for an extra training step and thus the need for additional labeled training data.

Finally, intermediate fusion is usually performed at kernel level. This can be done in different ways e.g. by taking the maximum of each kernel (which however, results in a non-Mercer kernel) or by computing the weighted sum:

$$K^* = \sum_{i=1}^{n} \omega_i * K_i \qquad (5)$$

where n is the number of different features and thus the number of different kernels. The kernel weights ω_i can be again optimized through grid-search and cross-validation or by more sophisticated techniques such as multi kernel learning (MKL, [11]).

Firstly, we combined the kernels of all spatial grid configurations using the weighted kernel sum fusion method. We set the kernel weights to $\omega_i = \frac{1}{3}$ meaning that each kernel is weighted equally. While this approach prevents us from optimizing another parameter it also means that concepts, which are represented better by a certain grid configuration are not ideally learned.

Next, we wanted to combine GIST and bag-of-words in order to investigate whether both features complement each other. The idea behind this concept was the assumption that the bag-of-words classifiers would perform better on object categories while GIST would perform better on scene categories. Again, we fused all 4 kernels using the weighted kernel sum and weighting the each kernel equally.

5   Experiments and Results

We applied our approach to the 2011 ImageClef Visual Concept Detection and Annotation Task's training dataset[3]. The task represents an extremely challenging image classification problem whose objective is to classify 10.000 photos taken from the Flickr community into 99 predefined classes. The authors provide a training set of 8000 images along with the ground truth label data. The image categories comprise concepts as diverse as "Partylife", "Building Sights", "Outdoor", "Lake", "Sunset/Sunrise", "Overexposed", "Out of focus", "MusicalInstrument", "Work", "natural", "dog", "skateboard", "male", "Adult" and "melancholic". While there are clearly a few concepts (e.g. "vehicle", "Flowers", "Trees", "River", "Food") that are more likely to be better annotated using local object-oriented classification methods such as bag-of-words there are also a number of concepts (e.g. "Outdoor", "Landscape& Nature", "Citylife", "Sunset Sunrise",

---

[3] ImageClef Visual Concept Detection and Annotation Task 2011 –
http://www.imageclef.org/2011/Photo

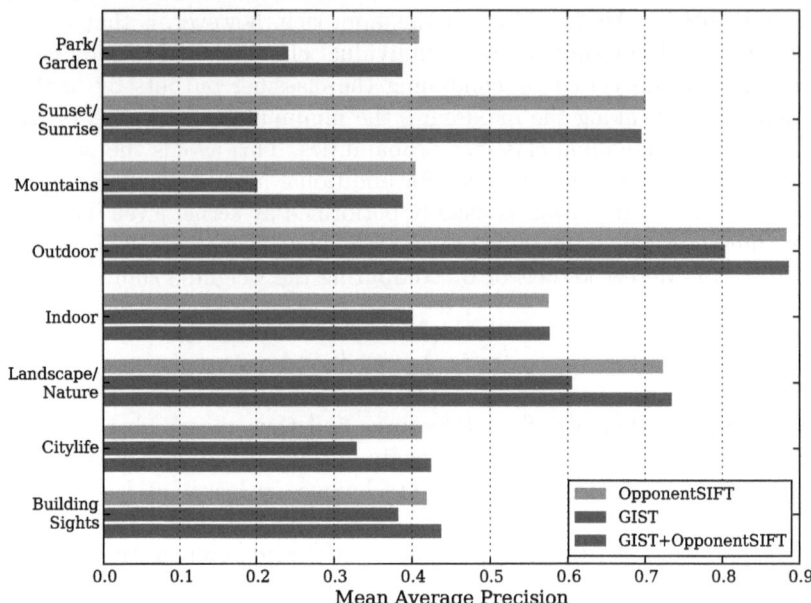

Fig. 3. Performance of GIST vs. OpponentSIFT on the ImageClef 2011 Visual Concept Detection and Annotation Task. The categories are sorted w.r.t. to the classification result of the OpponentSIFT classifier.

"Building Sights") that we assume are better classified using global techniques that better capture the spatial coherence such as GIST.

As we had no access to the testing data ground truth we split the training data into 5000 training and 3000 evaluation samples in order to be able to evaluate out approach. On the bag-of-words side we used the 5000 training samples to create the codebook and compute the training feature vectors. In order to compute the feature vectors for the evaluation samples we used the same codebook. We trained a Support Vector Machine on the 5000 feature vectors using the average kernel presented in 4.3. During training we only optimized the cost parameter C of the SVM in a 4-fold cross validation. Considering the GIST features we directly trained a SVM again by optimizing only the cost parameter by cross validation. We used both models to predict the labels of the 3000 samples in the evaluation set.

In order to visualize the performance of both approaches we subselected those categories of the overall 99 concept classes in the challenge that could be considered as scenic concepts: "Building Sights" (3), "Citylife"(5), "Landscape Nature"(6), "Indoor"(13),"Outdoor"(14),"Mountains"(24), "Sunset Sunrise"(28), "Park Garden"(53), Figure 3 shows the performance of the GIST and the OpponentSIFT classifiers on these concepts in terms of mean average precision (MAP) and the F-score per concept.

6   Conclusions

It is clearly visible is that OpponentSIFT outplays GIST in all categories. The mean average precision over all categories of OpponentSIFT is 29.5% whereas the mean average precision of GIST is 22.0%.

On the other hand, which becomes also evident is that the average performance of GIST for some categories is only slightly worse than the one of OpponentSIFT. This is interesting as the OpponentSIFT classifier demands for a computationally expensive feature vector extraction. Although codebook generation has to be performed only once during training, the extraction of OpponentSIFT features and the assignment of features to codewords has to be preformed for each image to be classified. GIST on the other hand can be extracted quite efficiently and there are a few categories (e.g. "Building Sights", "Outdoor" and "Buildings Sights") where it makes sense computationally-wise to replace OpponentSIFT by GIST as the prediction accuracy is only marginally worse. Therefore, the outcome of this evaluation is interesting as it shows that bag-of-words classification still is the yardstick by which all other approaches should be measured even for categories where global features seem to be more appropriate. However, global features should be considered as an alternative when it comes to computationally critical applications at the small price of slightly worse performance.

We also believe that the performance of the GIST classifier could be improved by a more carefully performed classifier optimization (e.g. by optimizing the kernel width as well). Moreover, instead of extraction GIST on gray level images and discarding color information there are likewise approaches to extract GIST in color space (e.g. see [3]) which should improve performance as well, however at the price of an increased feature dimensionality and therefore increased computational effort for distance computation.

Finally we wanted to see whether the fusion of GIST and OpponentSIFT can increase the overall performance and therefore combined both features at kernel level. We weighted all kernels equally and computed the kernel sum. Again we believe this cannot be expected to be the most optimal solution, however, it comes with reduced computational effort for weight optimization. The results are presented in Fig. 3, which shows that the combination does provide some improvements (averaged over all categories MAP increases by 0.04% when compared to OpponentSIFT alone). However, as can be seen, a performance increase can only be noticed if GIST alone performs at least half as good as OpponentSIFT alone.

A more thoroughly performed combination (e.g. by MKL or a late fusion approach) would most likely show better results here.

Acknowledgements. This work was supported in part by means of the Federal Ministry of Economics and Technology of Germany under the project THESEUS (01MQ07018).

References

1. Bosch, A., Zisserman, A., Munoz, X.: Representing shape with a spatial pyramid kernel. In: CIVR 2007: Proceedings of the 6th ACM International Conference on Image and Video Retrieval, pp. 401–408. ACM Press, New York (2007)
2. Csurka, G., Dance, C.R., Fan, L., Willamowski, J., Bray, C., Maupertuis, D.: Visual Categorization with Bags of Keypoints. In: Workshop on Statistical Learning in Computer Vision, ECCV, pp. 1–22 (2004)
3. Douze, M., Jégou, H., Sandhawalia, H., Amsaleg, L., Schmid, C.: Evaluation of GIST descriptors for web-scale image search. In: Proceeding of the ACM International Conference on Image and Video Retrieval, CIVR 2009, p. 1 (2009)
4. Friedman, A.: Framing pictures: The role of knowledge in automatized encoding and memory for gist. Journal of Experimental Psychology: General (1979)
5. Lazebnik, S., Schmid, C., Ponce, J.: Beyond Bags of Features: Spatial Pyramid Matching for Recognizing Natural Scene Categories. In: 2006 IEEE Computer Society Conference on Computer Vision and Pattern Recognition, CVPR 2006, vol. 2, pp. 2169–2178. IEEE (2006)
6. Leung, T., Malik, J.: Representing and Recognizing the Visual Appearance of Materials using Three-dimensional Textons. International Journal of Computer Vision 43(1), 29–44 (2001)
7. Lowe, D.G.: Distinctive Image Features from Scale-Invariant Keypoints. International Journal of Computer Vision 60(2), 91–110 (2004)
8. Oliva, A.: Gist of the Scene, ch. 41, pp. 251–257. Elsevier, San Diego (2005)
9. Oliva, A., Torralba, A.: Modeling the Shape of the Scene: A Holistic Representation of the Spatial Envelope. International Journal of Computer Vision 42(3), 145–175 (2001)
10. Snoek, C.G.M., Worring, M.: Concept-Based Video Retrieval. Foundations and Trends® in Information Retrieval 2(4), 215–322 (2009)
11. Sonnenburg, S., Rätsch, G., Schäer, C., Schölkopf, B.: Large scale multiple kernel learning. The Journal of Machine Learning Research 7, 1531–1565 (2006)
12. van De Sande, K.E., Gevers, T., Snoek, C.G.: A comparison of color features for visual concept classification. In: Proceedings of the 2008 International Conference on Content-Based Image and Video Retrieval, CIVR 2008, p. 141. ACM Press, New York (2008)
13. van de Sande, K.E.A., Gevers, T., Snoek, C.G.M.: Evaluating color descriptors for object and scene recognition. IEEE Transactions on Pattern Analysis and Machine Intelligence 32(9), 1582–1596 (2010)
14. Zhang, J., Marszałek, M., Lazebnik, S., Schmid, C.: Local Features and Kernels for Classification of Texture and Object Categories: A Comprehensive Study. International Journal of Computer Vision 73(2), 213–238 (2006)

Effectiveness of ICF Features for Collection-Specific CBIR

Nabeel Mohammed and David McG. Squire

Faculty of Information Technology, Monash University, Clayton Campus,
Wellington Road, Victoria 3800, Australia
{nabeel.mohammed,david.squire}@monash.edu

Abstract. This study aims to find more effective methods for collection-specific CBIR. A lot of work has been done in trying to adapt a system by user feedback, in this study we aim to adapt CBIR systems for specific image collections in an automated manner. Independent Component Analysis (ICA), a high order statistical technique, is used to extract Independent Component Filters (ICF) from image sets. As these filters are adapted to the data, the hypothesis is that they may provide features which are more effective for collection-specific CBIR. To test this question, this study develops a methodology to extract ICF from image sets and use them to extract filter responses. In developing this method, the study uses image cross-correlation and clustering to solve issues to do with shifted/duplicate filters and selecting a smaller set of filters to make CBIR practical. The method is used to generate filter responses for the VisTex database . The filter response energies are used as features in the GNU Image Finding Tool (GIFT). The experiments show that features extracted using ICF have the potential to improve the effectiveness of collection-specific CBIR, although some more work in this area is required.

1 Introduction

The primary aim of this study is to establish more effective techniques for Content Based Image Retrieval (CBIR) in collection-specific cases. With the wide availability of computing resources, a very large number of images are being produced, used and stored. However image searching has still not become nearly as effective or useful as text search [18]. This has to do with the various difficulties in searching for images, which can be based on either textual queries or image queries [16]. The latter is of interest to us in this study, as searching for images by using images as queries is the essence of CBIR.

While there have been successful results in using generic approaches (at times customised) for CBIR, the effectiveness of these systems can still be improved [15]. The current research laboratory prototypes of CBIR systems are still far behind from being available as effective commercial products [14]. As described in [16] it might be impossible to create a generic CBIR system (CBIRS) which performs well in all cases. A lot of research has taken place to improve CBIR performance based on user feedback. In this study we aim to explore a method to perform collection-specific CBIR, adapted for the collections, as it may be an area where great improvements are achievable.

M. Detyniecki et al. (Eds.): AMR 2011, LNCS 7836, pp. 83–95, 2013.

Independent Component Analysis (ICA), a high-order statistical technique, has been used with success to extract independent component filters (ICF) from images. These filters are are extracted in an unsupervised manner and are adapted to the images [12]. The research question is whether texture features extracted using these filters are more effective in CBIR compared to generic texture features. While there has been studies where ICA have been used to directly extract image features, to the knowledge of the authors no study has been carried out to establish whether the ICF extracted using ICA would lead to better features for CBIR. This study aims to answer this question. If the texture features extracted using ICF are shown to be more effective compared to generic approaches then this study would contribute new techniques to improve the effectiveness of collection-specific CBIR.

2   Background

2.1   Independent Component Analysis (ICA)

ICA is defined as "a method for finding underlying factors or components from multi-variate (multidimensional) statistical data" [9]. The initial motivation behind ICA was to perform Blind Source Separation (BSS), which refers to the task of discovering the source signals from some observed linear mixture of the sources [9]. In fact BSS is a good example to describe ICA as a mathematical problem. Here a simple version of ICA is presented, where we assume that the number of observed signals and the number of source signals are equal. Let $x_1(t)$, $x_2(t)$ and $x_3(t)$ represent the observed signals of some source signals $s_1(t)$, $s_2(t)$ and $s_3(t)$ at time t. Based on this information, it can be said that for $i = 1, 2, 3$

$$x_i(t) = a_{i1}s_1(t) + a_{i2}s_2(t) + a_{i3}s_3(t). \tag{1}$$

In this situation, the source signals $s_i(t)$ and the mixing weights a_{ij} are unknown. The only known values are the observed signals x_{ij}. The problem of BSS is to find the original signals $(s_i(t))$ from the observed mixtures $(x_i(t))$. The assumption is that there is an invertible matrix A formed from the mixing coefficients a_{ij}. The inference then is, there is a matrix W, with w_{ij} as coefficients, which would allow the separation of each s_i, as

$$s_i(t) = w_{i1}x_1(t) + w_{i2}x_2(t) + w_{i3}x_3(t). \tag{2}$$

That is, $W = A^{-1}$. This is the basic mathematical problem. ICA provides a solution to this seemingly hard problem by the assuming that the signals are statistically independent. That is, if v_1 and v_2 are independent, then for any non-linear transformations f and g, $f(v_1)$ and $g(v_2)$ will also be uncorrelated [9]. So, in essence the task of ICA is to find the components such that the components themselves are uncorrelated and also remain uncorrelated under non-linear transformations f and g.

An important principle for estimating independent components is the maximisation of non-Gaussianity. The central limit theorem states that the sum of non-Gaussian random variables will be closer to a Gaussian compared to the original ones. Therefore finding maxima in non-Gaussianity in a linear combination of the mixture variables $(y = \sum_i b_i x_i)$ gives us the independent components [9].

2.2   ICA in CBIR

As ICA finds underlying components from a dataset and has been applied to images, various studies have attempted to use it for CBIR. [11] compares ICs extracted from the query image and images in the database to determine the results. The paper mentions the use of a ICA filter bank but does not clarify how the filters were designed. It seems as if the process proposed extracts ICs from the query image, uses them as filters and collects filter responses from the database. If this is indeed the case, then there are certain issues with the approach. Firstly, it requires repetitive execution of ICA on the query image. This can be quite an expensive process. Also, ICA can extract a large number of components and it is important to reduce this number for practical CBIR. [19] also uses ICA in conjunction with Generalized Gaussian Density for the purposes of CBIR. The results shown in the paper are very encouraging, however their method also suffers from the use of ICA in the image feature extraction process. [1] uses Probabilistic ICA to extract image features and uses the z-values of ICs to find a component-wise similarity bipartite. [21] uses ICA features and other low-level features, along with a learning algorithm for image retrieval and they show very promising results. Although not directly related to CBIR, there has been some work done in applying ICA for image features for a variety of tasks, including segmentation, classification, dimensionality reduction, etc. [8] [13] [17] [19] [22].

2.3   Our Approach

In image processing, ICA can be used to extract components from sets of images. These components can be transformed back into patches to form filters. Each of the filters is known as an Independent Component Filter (ICF) [6]. Hateren et al. [6] state that when ICA is applied to images of natural scenes, it produces components similar to the receptive fields in simple cells in the visual cortex. [2] describe the use of ICA to extract filters from images of natural scenes, and say those filters are edge filters, noting their resemblance to Gabor filters.

In the studies mentioned in 2.2, ICA has been used to extract image features which were then used to perform CBIR. This study attempts to use techniques from work such as [6] and [2] and use it in the context of CBIR. This differentiates our work from that of [11], [19] etc. The advantage of this approach is that ICA only needs to be executed once on each image set, at the time of learning the filters. Once the filters have been found, we can extract features from the images using filter responses. This should be faster and more scalable for use in production quality systems.

3   Experiments

This study used a modified version of the VisTex database[1]. It is a database holding a fixed set of images of various kinds of texture. Some example images are shown in Figure 1. The original database had 512×512 images. The version used for this study,

---

[1] Vistex Database is available from
  http://vismod.media.mit.edu/vismod/imagery/VisionTexture/

uses slightly modified version of the original, which was first used in the development of the Viper/GIFT system at the University of Geneva [18]. Ten 256×256 patches were taken randomly from the images of the database and downsized to 128×128. Using this version allowed the study in [18] to proceed from an established ground-truth, as all the patches taken from a single image can be taken to be similar, specially considering that most images in the VisTex database have a uniform texture throughout, as can be seen from Figure 1.

Fig. 1. 3 images from the VisTex database

However, as some of the original images in the VisTex database were very similar, the relevance judgement based on sub-images did not seem to be accurate. To address the problem, this study carried out relevance judgements for 10 sample query images to be used in the experiments. For all the images in the database, three sets of features were extracted. They are explained here.

3.1   Feature Sets

Existing GIFT Features Using a Bank of Gabor Filters. A complete explanation of the feature types of GIFT can be found in [18]. Here we will present information related only to the texture features and only the information deemed to be most relevant.

GIFT employs a bank of real, circularly symmetric Gabors:

$$f_{mn}(x, y) = \frac{1}{2\pi\sigma_m^2} e^{-\frac{x^2+y^2}{2\sigma_m^2}} \cos(2\pi(u_{0_m}x \cos\theta_n + u_{0_m}y \sin\theta_n)), \qquad (3)$$

where m indexes filter scales, n their orientations, and u_{0_m} is the centre frequency. The half peak radial bandwidth is chosen to be one octave, which determines σ_m. The highest centre frequency is chosen as $u_{0_1} = 0.5$, and $u_{0_{m+1}} = u_{0_m}/2$. Three scales are used. The four orientations are: $\theta_0 = 0$, $\theta_{n+1} = \theta_n + \pi/4$. The resultant bank of 12 filters gives good coverage of the frequency domain, with little filter overlap. The mean energy of each filter is computed for each of the smallest blocks in the image, and quantized into 10 bands. A feature is stored for each filter with energy greater than the lowest band. These are treated as local features. Each image has at most 3072 of the 27648 such possible features. Histograms of these features are used to represent global texture characteristics.

ICF Features. For filter extraction, one image was chosen to represent each texture class. We implemented the FastICA algorithm [9], and applied it to these selected images. Experimentation was done with various different patch sizes, however this paper only presents the results gathered from 17×17 patches. $10,000$ random patches were extracted from the training set. For some of the experiments, a 17×17 Hamming window was applied on each patch. However a normal $2D$ Hamming window has a sharp rise and a narrow peak, leading to massive reduction in dimensionality when using PCA. In an attempt to get a slightly wider peak, the window used in this study is an elementwise square root of the original Hamming window. The use of a Hamming window has been recommended in literature [3], but there is no evidence showing whether ICFs extracted from image patches which had a Hamming window applied to it, were more effective than ICFs which were extracted from unprocessed image patches. Although it is not the main goal of this study to perform such a comparison, this study will provide some insight about the usefulness of Hamming windows for patch processing. For the purpose of this paper, the features extracted using ICFs with the Hamming window shall be called HammingICF features and the features extracted using ICFs without the Hamming window shall be called the ICF features.

As the patch sizes used in this study is 17×17 the original dimension was 289. PCA was used to drop the least significant 1% of the dimensions before applying ICA. Table 1 shows the dimensionality after PCA and the number of ICFs extracted.

Table 1. New dimensions after running PCA and number of ICs

Hamming Window	New dimensions	No. of IC
Yes	180	110
No	231	146

[20] mentions the problem where ICA extracts ICFs which are apparently shifted or near duplicate versions of each other. As these filters would extract very similar features, it is preferable to eradicate the shifted/duplicate versions. Also, ICA extracts very large number of ICFs. For practical CBIR, we need to select a smaller subset of filters. [4] uses cross-correlation with Self Organising Maps (SOMs) to select filters, but provides no motivation for using SOMs for this problem—there is no apparent reason to expect the ICFs to lie on a two-dimensional manifold.

This study has employed a similar but simpler approach. For each pair of filters a cross-correlation matrix was calculated. From each of these matrices the highest value was chosen as the cross-correlation value of the two filters. Using these values a matrix of correlation values of each pair of filters was constructed. This matrix was used as a similarity measure in an implementation of complete-link clustering [10]. To find filters to use in CBIR, a decision was made on how many filters we want to use for feature extraction. The threshold value given to the clustering algorithm was adjusted appropriately to return the desired number of filters. As an example, when about 20 filters were required, a threshold value of 0.14 was used giving 22 filters. This is one crucial area of this study and needs further research to establish the validity of this technique. However a visual inspection was conducted and the clustering technique seemed to be grouping filters with similar visual layout. An example of a cluster is shown in Figure 2. The filters have been scaled for visual presentation.

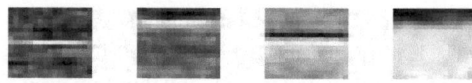

Fig. 2. Complete link clustering grouped these filters in a cluster when a threshold of 0.14 was used

From each cluster, the filter which has the highest average correlation with other filters in the same cluster is chosen. The chosen filters were used in $2D$ convolution and GIFT used the filter energies to extract image features using the approach mentioned for the bank of Gabor filters.

GLCM Based Features. As both the Gabor features and the ICF features are extracted using filters, we wanted to compare their performance with other types of texture features. One of the most common and intuitive method for texture feature extraction is using Grey Level Co-occurrence matrices. [5] suggested the use of GLCMs and provided a good set of features which can be extracted from the GLCMs of images. A GLCM conveys information about the frequency of two grey level values, separated by a certain vector, appearing in an image. Changing the angle and size of the vector will give different GLCMs. There have been many different methods of applying GLCMs, but we chose to implement the method used by [7]. The work in [7] was carried out for medical images, which could be classified as a specialised collection.

The study in [7] divided each images in 7×7 tiles and for each tile calculated 16 GLCMS. The GLCMS were generated for vectors of size 1,2,3 and 4 pixels and orientations 0, $\frac{\pi}{4}$, $\frac{\pi}{2}$ and $\frac{3\pi}{4}$. For each GLCM $P(i, j)$, they calculated a homogeneity feature H_p,

$$H_p = \sum_i \sum_j \frac{P(i, j)}{1 + |i - j|} \tag{4}$$

Using these features they calculated the Manhattan distance between the query images and the images in the database.

For our work, we implemented their scheme and used it as local GLCM features. For global GLCM features, we calculated GLCMs for whole images, rather than breaking it up into tiles. When combining global and local features, we followed the recommendation in [7] and divided distances of each feature by their median and then summing up the values.

3.2   Setup and Feature Sets

GIFT as a CBIRS has a unique property that it uses MRML, an XML-based communication protocol, to communicate between the CBIR client and the CBIRS and also between its own components[2]. This allowed the development of a suite of scripts which can be used to automatically run tests and gather the results.

---

[2] http://www.mrml.net/

For the tests, 10 query images were selected from the image database. The existing features, extracted using a bank of Gabor filters, performed adequately on some of these images but not very well on most of them. Example of such images are shown in figure 3.

(a)   (b)

Fig. 3. Two example query images. The Gabor features perform well on (a) and poorly on (b).

Precision-Recall graphs were generated based on the relevance judgements and used to compare CBIR performance for features extracted using three different filter collections and the GLCMs, giving the following feature sets.

- Features extracted using a bank of 12 Gabor filters (Gabor features).
- Features extracted using ICF without Hamming window (ICF features).
- Features extracted using ICF with Hamming window (HammingICF features).
- Features extracted using GLCM features (GLCM Features).

Experiments were conducted to find the CBIR performance for global features, local features and using both global and local features and the results are described next.

4   Results and Discussion

Figures 4, 5 and 6 shows the Precision vs Recall graphs generated from the experiment results for the VisTex database, where precision is average over all query images. As can be seen from the graphs, the ICF features have overall better performance compared to the Gabor features. The HammingICF features perform better than the Gabor features, however not as well as the ICF features. This is a bit surprising, as it was theorised that eliminating hard edges in the filters would produce more useful results, but the use of the Hamming window seems hamper the extraction of useful image features. When using local features only, the performance of the GLCM features is clearly superior to the filter based features, although not close to the global features.

For a query image, the GIFT interface shows 20 images that GIFT deems to be most relevant depending on the features used. For each query image, the precision in the first 20 results were calculated. These results are shown in Figures 7, 8 and 9. It is clear from these graphs that for certain image classes the Gabor features give good results, but for a majority of the query images the ICF feature and the HammingICF features outperform the Gabor and GLCM features . This measurement of performance is important as it can be directly translated to a better user experience.

As is evident from the results, the global features outperform the local features. Using the combination of global and local features actually seems to worsen performance.

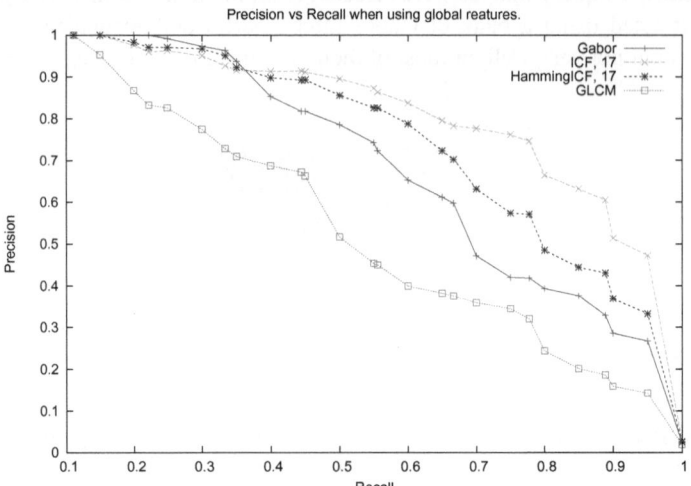

Fig. 4. Precision vs Recall graph for queries using global features

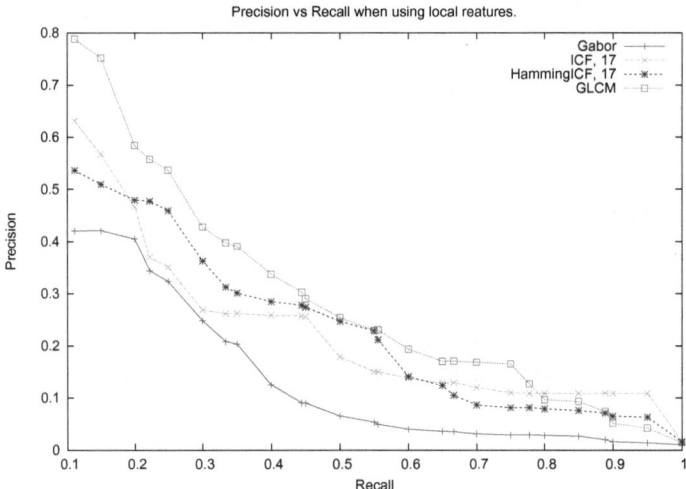

Fig. 5. Precision vs Recall graph for queries using local features

This can be explained by the nature of the images in the VisTex database. As mentioned previously, most of the images have a uniform texture; it is reasonable that features which express image characteristics as whole would work better. Local feature which would identify differences in image regions might actually be counter-productive for a database of images such as VisTex. The only exception to this seems to be the performance of the GLCM features when querying with images of buildings. The combination of local and global GLCM features gives perfect result. This needs further study to

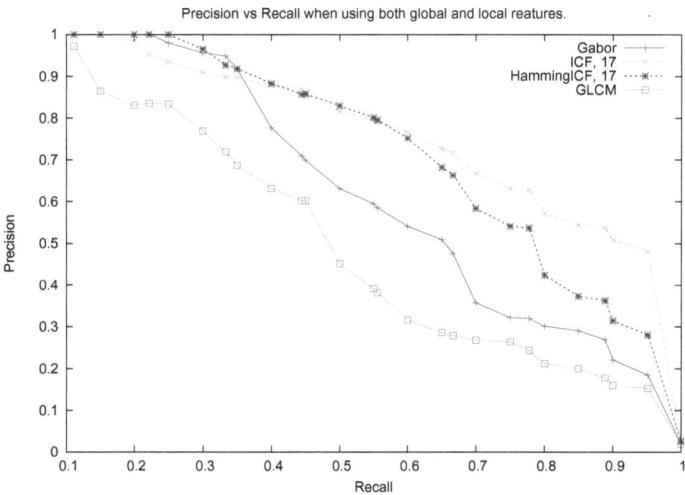

Fig. 6. Precision vs Recall graph for queries using both global and local features

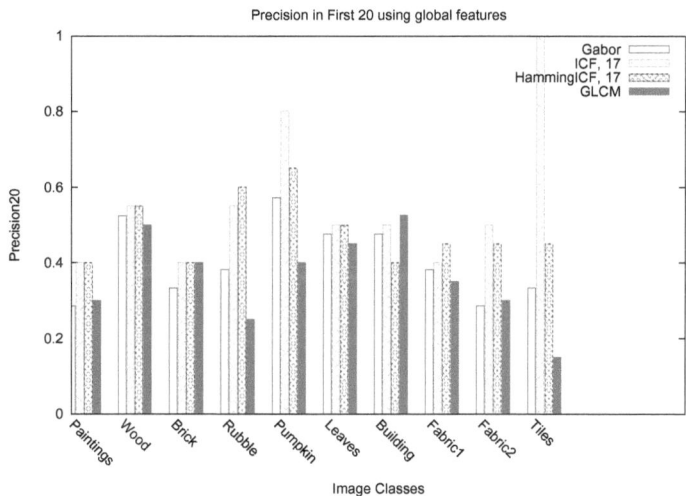

Fig. 7. Precision for different image classes in the top 20 results, using Global features only

ascertain whether for certain types of images the GLCM features are able to capture more useful information.

It was theorised that ICF extracted using ICA would be adapted to the data and hence would help in finding texture features to which the Gabor filters and other pre-determined methods would be blind. An example of such a case is shown in the two query images in Figures 10(a) and 10(b). Both of them are images of very similar fabrics. Figure 10(a) shows a fabric which has larger texture compared to the fabric shown

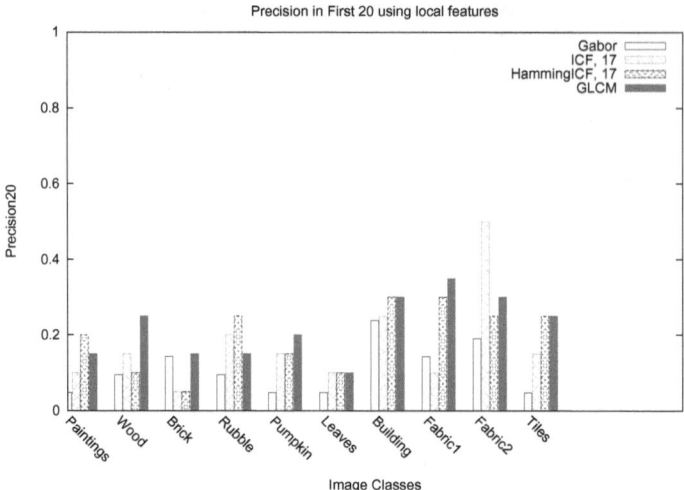

Fig. 8. Precision for different image classes in the top 20 results, using Local features only

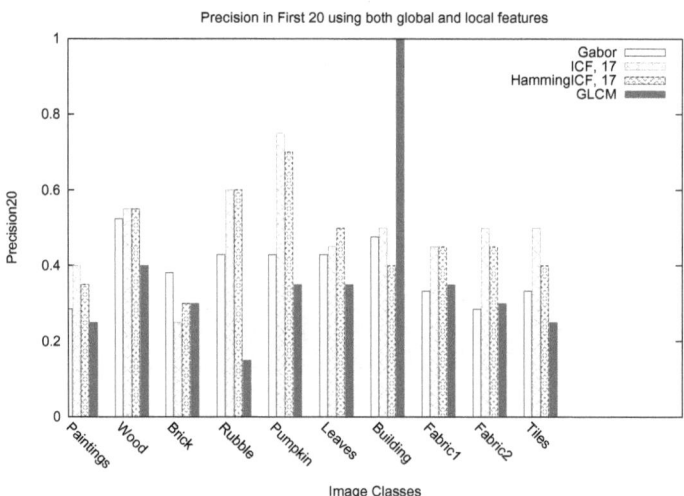

Fig. 9. Precision for different image classes in the top 20 results, using both local and global features

in Figure 10(b). When using Gabor features, the first query image (Figure 10(a)) has a precision value of 0.4 in the first 10 images retrieved compared to a precision value of 0.5 for the GLCM features and 1 for the ICF-based features. On the other hand for Figure 10(b) the Gabor, GLCM and ICF-based features, all have a precision value of 0.7 for the first 10 images retrieved. It is also interesting to note that for Figure 10(a) when using Gabor features, images of water are judged to be similar and when using

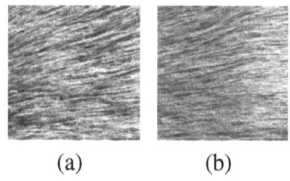

(a) (b)

Fig. 10. Two very similar query images. For (a) the Gabor features perform poorly, for (b) they perform well.

GLCM features, images of powdered food is judged to be similar. This is an example where the pre-determined size and orientation of the bank of Gabor filters and GLCMs leads to inaccurate results.

5 Conclusion

The results indicate that for images with globally consistent texture features, the ICF features work well. Analysing the precision at the top 20 results, the global ICF-based features seem to perform better than both the Gabor and GLCM features. However for a single image (buildings), the global GLCM features perform better. The ICF-based features seem to describe image characteristics which the Gabor and GLCM features have been unable to capture. This is a strong indication that learning filters from image sets and using those filters to extract features can be a viable way to perform CBIR feature extraction, without incurring the cost of having the overhead of running ICA repeatedly.

Extracting the ICF is more computationally expensive than using Gabor filters. Further studies need to explore how this process can be made less expensive. There are other areas which may give ICF features better performance, crucial among those are choosing the filters to use in CBIR. This study uses cross-correlation followed by clustering, however other methods need to be explored to ensure that the optimal set of filters is chosen.

It is also acknowledged that this study has the same limitations as other CBIR studies. The results are based on the image database chosen and also the relevance judgements made. Work is currently under progress to implement similar processes across different image collections, including a set of images with varying texture within the image and also one of skin lesions. Also, an effort is being made to establish a comprehensive relevance judgement set for the VisTex database and the other databases. The results of this study gives encouragement to pursue this further work.

References

1. Bai, B., Kantor, P., Shokoufandeh, A., Silver, D.: fMRI brain image retrieval based on ICA components. In: Eighth Mexican International Conference on Current Trends in Computer Science, ENC 2007, September 24-28, pp. 10–17 (2007)
2. Bell, A.J., Sejnowski, T.J.: The "independent components" of natural scenes are edge filters. Vision Research 37(23), 3327–3338 (1997),
   http://dx.doi.org/10.1016/S0042-69899700121-1

3. Borgne, H.L., Guerin-Dugue, A.: Sparse-dispersed coding and images discrimination with independent component analysis. In: Proceedings of the Third International Conference on Independent Component Analysis and Signal Separation, ICA 2001 (December 2001)

4. Guo, C., Wilson, C.: Use of self-organizing maps for texture feature selection in content-based image retrieval. In: IEEE International Joint Conference on Neural Networks, pp. 765–770 (June 2008)

5. Haralick, R.M., Shanmugam, K., Dinstein, I.: Textural features for image classification. IEEE Transactions on Systems, Man and Cybernetics 3(6), 610–621 (1973), http://dx.doi.org/10.1109/TSMC.1973.4309314

6. Hateren, J.H.V., Schaaf, A.V.D.: Independent component filters of natural images compared with simple cells in primary visual cortex. In: Proceedings of the Biological sciences/The Royal Society, vol. 265(1394) (1998)

7. Howarth, P., Yavlinsky, A., Heesch, D., Rüger, S.M.: Medical image retrieval using texture, locality and colour. In: Peters, C., Clough, P., Gonzalo, J., Jones, G.J.F., Kluck, M., Magnini, B. (eds.) CLEF 2004. LNCS, vol. 3491, pp. 740–749. Springer, Heidelberg (2005), http://dx.doi.org/10.1007/1151964572,

8. Huang, B., Li, J., Hu, S.: Texture feature extraction using ICA filters. In: 7th World Congress on Intelligent Control and Automation, WCICA 2008, pp. 7631–7634 (June 2008)

9. Hyvarinen, A., Karhunen, J., Oja, E.: Independent Component Analysis. Wiley-Interscience (May 2001), http://www.amazon.ca/exec/obidos/redirect?tag=citeulike09-20&path=ASIN/047140540X

10. Jain, A.K., Murty, M.N., Flynn, P.J.: Data clustering: a review. ACM Comput. Surv. 31(3), 264–323 (1999), http://dx.doi.org/10.1145/331499.331504

11. Khaparde, A., Deekshatulu, B.L., Madhavilatha, M., Farheen, Z., Sandhya Kumari, V.: Content based image retrieval using independent component analysis. IJCSNS International Journal of Computer Science and Network Security 8(4), 327–332 (2008)

12. Le Borgne, H., Guérin-Dugué, A., Antoniadis, A.: Representation of images for classification with independent features. Pattern Recognition Letters 25(2), 141–154 (2004), http://dx.doi.org/10.1016/j.patrec.2003.09.011

13. Li, Z., Wu, S., Wang, X., Ye, H., Wang, M., Ye, J.: Dimensional reduction based on independent component analysis for content based image retrieval. In: International Joint Conference on Artificial Intelligence, JCAI 2009, pp. 741–745 (2009)

14. Marques, O., Furht, B.: Content-Based Image and Video Retrieval. Kluwer Academic Publishers, Norwell (2002)

15. Müller, H., Rosset, A., Vallee, J., Geissbuhler, A.: Comparing feature sets for content based image retrieval in a medical case database. In: Proceedings of the Medical Informatics Europe Conference, MIE 2003, St. Malo, France (May 2003)

16. Müller, W.: Design and implementation of a flexible Content–Based Image Retrieval Framework - The GNU Image Finding Tool. Ph.D. thesis, Computer Vision and Multimedia Laboratory, University of Geneva, Geneva, Switzerland (September 2001)

17. Smoaca, A., Coltuc, D., Lazarescu, V.: Pattern segmentation in textile images. In: International Symposium on Signals, Circuits and Systems, ISSCS 2009, pp. 1–4 (July 2009)

18. Squire, D.M., Müller, W., Müller, H., Raki, J.: Content-based query of image databases, inspirations from text retrieval: inverted files, frequency-based weights and relevance feedback. Pattern Recognition Letters, 143–149 (1999)

19. Sun, G., Liu, J., Sun, J., Ba, S.: Locally salient feature extraction using ICA for content-based face image retrieval. In: International Conference on Innovative Computing, Information and Control, vol. 1, pp. 644–647 (2006)
20. Trojan, N.: CBIR-based Dermatology Diagnostic Assistant. Honours thesis, School of Computer Science and Software Engineering, Monash University, Clayton, Victoria, Australia (November 2004); supervised by Dr. David Squire
21. Wang, F., Dai, Q.-H.: A new multi-view learning algorithm based on ICA feature for image retrieval. In: Cham, T.-J., Cai, J., Dorai, C., Rajan, D., Chua, T.-S., Chia, L.-T. (eds.) MMM 2007. LNCS, vol. 4351, pp. 450–461. Springer, Heidelberg (2006)
22. Zhang, R., Zhang, X.P., Guan, L.: Wavelet-based texture retrieval using independent component analysis, vol. 6, pp. VI-341–VI-344 (September 2007)

An Experimental Comparison of Similarity Adaptation Approaches

Sebastian Stober and Andreas Nürnberger

Data & Knowledge Engineering Group
Faculty of Computer Science
Otto-von-Guericke-University Magdeburg, D-39106 Magdeburg, Germany
{Sebastian.Stober,Andreas.Nuernberger}@ovgu.de

Abstract. Similarity plays an important role in many multimedia retrieval applications. However, it often has many facets and its perception
is highly subjective – very much depending on a person's background or
retrieval goal. In previous work, we have developed various approaches
for modeling and learning individual distance measures as a weighted
linear combination of multiple facets in different application scenarios.
Based on a generalized view of these approaches as an optimization problem guided by generic relative distance constraints, we describe ways to
address the problem of constraint violations and finally compare the
different approaches against each other. To this end, a comprehensive
experiment using the *Magnatagatune* benchmark dataset is conducted.

1 Introduction

Similarity or distance measures are a crucial component in any information retrieval system in general. Particularly in multimedia retrieval, the objects of
consideration can often be compared w.r.t. a multitude of facets. For instance,
the distance of two music pieces could be computed based on the rhythm, tempo,
lyrics, melody, harmonics, timbre or even mood. While sophisticated measures
usually exist for each single facet, the question remains how to obtain a suitable
combination of multiple facets that reflects the background and the retrieval goal
of the user. In previous work, we have proposed various approaches for adapting
individual distance measures as a weighted linear combination of multiple facets
and demonstrated how they can be applied in several real-world interactive scenarios: organizing and exploring the work of the Beatles with the *BeatlesExplorer*
user interface by re-arranging songs and correcting rankings [17], learning suitable similarity measures for folk song classification from expert annotations [2],
and tagging photographs [18]. Each time, a different learning approach was taken
to obtain the desired adaptation with different application-depending objectives.
However, as described recently [19] and recapitulated in Section 2, there is a generalized view which is also consistent with various related works (Section 3). This
formalization provides a unified model for our adaptation approaches taken so
far which are briefly outlined in Sections 4.1 to 4.3. Additionally, we address the

M. Detyniecki et al. (Eds.): AMR 2011, LNCS 7836, pp. 96–113, 2013.
© Springer-Verlag Berlin Heidelberg 2013

problem of constraint inconsistencies in Section 4.4 and furthermore propose alternative problem formalizations that allow constraint violations in Section 4.5. As the main contribution of this paper, Section 5 describes the experimental comparison of the approaches covered by Section 4 using the *Magnatagatune* benchmark dataset. Finally, Section 6 concludes this paper.

2   Formalization

To begin with, the concept of facet distances needs to be formalized assuming a feature-based representation of the objects of interest:

Definition 1. *Given a set of features F, let S be the space determined by the feature values for a set of objects O. A facet f is defined by a facet distance measure δ_f on a subspace $S_f \subseteq S$ of the feature space, where δ_f satisfies the following conditions for any $a, b \in O$:*

- *$\delta_f(a, b) \geq 0$ and $\delta_f(a, b) = 0$ if and only if $a = b$*
- *$\delta_f(a, b) = \delta_f(b, a)$ (symmetry)*

Furthermore, δ_f is a distance metric if it additionally obeys the triangle inequality for any $a, b, c \in O$:

- *$\delta_f(a, c) \leq \delta(a, b) + \delta(b, c)$ (triangle inequality)*

In order to avoid a bias when aggregating several facet distance measures, the values need to be normalized. The following normalization is applied for all distance values $\delta_f(a, b)$ of a facet f:

$$\delta'_f(a, b) = \frac{\delta_f(a, b)}{\mu_f} \tag{1}$$

where μ_f is the mean facet distance with respect to f:

$$\mu_f = \frac{1}{|\{(a, b) \in O^2\}|} \sum_{(a,b) \in O^2} \delta_f(a, b) \tag{2}$$

As a result, all facet distances have a mean value of 1.0. Special care has to be taken, if extremely high facet distance values are present that express "infinite dissimilarity" or "no similarity at all". Such values introduce a strong bias for the mean of the facet distance and thus should be ignored during its computation.

The actual distance between objects $a, b \in O$ w.r.t. the facets $f_1, \ldots, f_l$ is computed as weighted sum of the individual facet distances $\delta_{f_1}(a, b), \ldots, \delta_{f_l}(a, b)$:

$$d(a, b) = \sum_{i=1}^{l} w_i \delta_{f_i}(a, b) \tag{3}$$

This way, facet weights $w_1, \ldots, w_l \in \Re$ are introduced that allow to adapt the importance of each facet according to user preferences or for a specific retrieval task. These weights obviously have to be non-negative and should also have an upper bound, thus:

$$w_i \geq 0 \qquad \forall 1 \leq i \leq l \tag{4}$$

$$\sum_{i=1}^{l} w_i = l \tag{5}$$

They can either be specified manually or learned from preference information. In the scope of this work, all preference information is reduced to *relative distance constraints*.

Definition 2. *A relative distance constraint* (s, a, b) *demands that object* a *is closer to the seed object* s *than object* b, *i. e.:*

$$d(s, a) < d(s, b) \tag{6}$$

With Equation 3 this can be rewritten as:

$$\sum_{i=1}^{l} w_i(\delta_{f_i}(s, b) - \delta_{f_i}(s, a)) = \sum_{i=1}^{l} w_i x_i > 0 \tag{7}$$

substituting $x_i = \delta_{f_i}(s, b) - \delta_{f_i}(s, a)$. Such basic constraints can directly be used to guide an optimization algorithm that aims to identify weights that violate as few constraints as possible [16]. Alternatively, the positive examples $(x, +1)$ and the negative examples $(-x, -1)$ can be used to train a binary classifier in which case the weights $w_1, \ldots, w_l$ define the model (separating hyperplane) of the classification as pointed in [4]. Furthermore, the relative distance constraints are still rich enough to cover more complex forms of preference as summarized in [19]. Also note that such relative statements are usually much easier to formulate than absolute ones.

3   Relation to Other Approaches

An adaptive structuring technique for text and image collections using Self-Organizing Maps (SOMs) is described in [13]. The underlying weighted Euclidean distance is automatically adapted according to user feedback (changing the location of objects in the SOM). For the adaptation, no distance constraints as described by Equation 6 are used explicitly. Instead, weight update rules are applied based on the feature differences which correspond to the x_i in Equation 7.

Bade [1] applies metric learning for personalized hierarchical structuring of (text) collections where each document is represented by a vector of term weights [14]. To this end, structuring preferences are modeled by so-called *"must-link-before constraints"* – each referring to a triple (a, b, c) of documents. Such a constraint expresses a relative relationship according to hierarchy levels, namely that a and b should be linked on a lower hierarchy level than a and c. In hierarchical clustering, this means nothing else but that a and b are more similar than a and c. Consequently, *must-link-before constraints* can be considered as a domain-specific interpretation of the more generic relative distance constraints (Equation 6).

Cheng et al. [4] approach the metric learning problem from a case-based reasoning perspective and thus call the relative distance constraints *"case triples"*. In contrast to the previously outlined works, they address object representations beyond plain feature vectors: They model similarity as a weighted linear combination of *"local distance measures"* which corresponds to the facet concept and the aggregation function used here. As a major contribution, they also show how this formulation of the metric learning problem can be interpreted as a binary classification problem that can be solved by efficient learning algorithms and furthermore allows non-linear extensions by kernels.

The work of McFee et al. [12,11] on Metric Learning to Rank (MLR) – an extension of the Structural Support Vector Machine (SVM) approach [8] – is related in that their learning methods are also guided by relative distance constraints (which they call *"partial order constraints"*). They also combine features from different domains (acoustic, auto-tags and tags given by users) which could be interpreted as facets with the respective kernels corresponding to facet similarity measures. However, their approaches differ from the one addressed in this paper in that they aim to learn an embedding of the features into an Euclidean space and to this end apply complex non-linear transformations using kernels. Whilst their techniques are more powerful in the sense that they allow to model complex correlations, this comes at a high price: The high complexity is problematic when users want to understand or even manually adapt a learned distance measure. Here, the simplicity of the linear combination approach is highly beneficial.

Wolff and Weyde [20] also apply MLR for learning a Mahalanobis distance that reflects a perceived or stated music similarity according to relative similarity ratings by users. Their experiments are based on the *Magnatagatune* dataset, which is also used here. However, they additionally incorporate genre tags of the corresponding albums from *Magnatune* as features and furthermore take a different approach to derive constraints (referred to as *"binary rankings"*) from the similarity judgments (cf. Section 5.2). Therefore, the results cannot be compared directly. In general, a higher number of satisfiable constraints can be expected as the Mahalanobis distance has more degrees of freedom for adaptation (a $n \times n$ matrix of covariances as opposed to n facet weights), However, the convergence behavior of the incremental adaptation remains unclear. For a meaningful comparison, both, the features and the evaluation methodology, need to be identical.

Slaney et al. [15] state that they *"use labels* [artist, label and blog] *to tune the Mahalanobis matrix so that similar songs are likely to be close to each other in the metric space."* Although this is not explicitly stated in their description, this also implies either relative distance constraints (as used here) or absolute constraints of the form *"Songs a and b have to be in the same cluster."* However, using the Mahalanobis distance, they require more restrictive plain vector representations as input. Furthermore, the resulting music similarity model – i.e., the covariance matrix of the Mahalanobis distance – is still harder to interpret and adapt manually.

Finally, the *SoniXplorer* [10] is in many aspects similar to the *BeatlesExplorer* prototype described earlier in [17]: The system covers multiple facets and uses a weighted linear aggregation for the underlying similarity measure. However, the adaptation is not guided by relative distance constraints. Instead, the system allows the users to specify distance information by manipulation of the terrain, i. e., the formation of new separating hills or their removal respectively. By numerical integration over the height profile, a target distance matrix for the learning algorithm is derived that contains *absolute* (quantitative) distance information.

4   Optimization Approaches

It is possible to look at the optimization problem introduced in Section 2 from different perspectives: tolerance w.r.t. constraint inconsistencies, stability, continuity and responsiveness. The following sections describe three different optimization approaches that have been taken in previous work – each one for a different application: a gradient descent approach (Section 4.1), a Quadratic Programming (QP) approach (Section 4.2), and a maximum margin approach (Section 4.3). Furthermore, Section 4.4 describes a generic way of dealing with constraint inconsistencies and Section 4.5 proposes several alternative QP problem formulations that allow constraint violations.

4.1   Gradient Descent

In the folk song classification experiments described in [2], weights are learned by a gradient descent approach similar to the work in [3]. During learning, all constraint triples (s, a, b) are presented to the algorithm several times until convergence is reached. If a constraint is violated by the current distance measure, the weighting is updated by trying to maximize

$$obj(s, a, b) = \sum_{i=1}^{l} w_i (\delta_{f_i}(s, b) - \delta_{f_i}(s, a)) \tag{8}$$

which can be directly derived from Equation 7. This leads to the update rule for the individual weights:

$$w_i = w_i + \eta \Delta w_i \tag{9}$$

$$\text{with} \quad \Delta w_i = \frac{\partial obj(s, a, b)}{\partial w_i} = \delta_{f_i}(s, b) - \delta_{f_i}(s, a) \tag{10}$$

where the learning rate η defines the step width of each iteration.[1] To enforce the bounds on w_i given by Equation 4 and Equation 5, an additional step is necessary after the update, in which all negative weights are set to 0 and then the

---

[1] Approaching the weight learning problem from the classification perspective using a perceptron for classification as described in [4] results in the same update rule.

weights are normalized to sum up to l. This algorithm can compute a weighting, even if not all constraints can be satisfied due to inconsistencies. However, no largest margin is enforced. Using the current weights as initial values in combination with a small learning rate allows for some continuity but there may still be solutions with less change required. It is possible to limit the number of iterations to increase responsiveness but this may result in some unsatisfied constraints.

4.2 Quadratic Programming: Minimizing Weight Change

For maximum continuity which is considered most important in the *BeatlesExplorer* application described in [17], the weights should change only as little as necessary to satisfy all constraints. This can directly be modeled as a Quadratic Programming (QP) problem demanding in the objective function that the sum over all (quadratic) deviations of the weights from their previous values should be minimal (with initial values 1):

$$\min_{(w_1,\ldots,w_l)\in\Re^l} \sum_{i=1}^{l} \left(w_i - w_i^{(old)}\right)^2 \tag{11}$$

subject to the constraints that enforce the weight bounds (Equation 4 and Equation 5) and the distance constraints (Equation 7) which can be used directly. The problem can be solved using the Goldfarb and Idnani dual QP algorithm for convex QP problems subject to general linear equality/inequality constraints [7]. For this original formalization of the weight learning problem, there is only a solution if all constraints are consistent. Section 4.5 proposes different ways to integrate slack variables which allow the violation of constraints.

4.3 Maximal Margin Classifier

If stability is more important than continuity, the primary objective is to maximize the margin between the separating hyperplane and the positive and negative training samples (generated from the distance constraints as described in Equation 7). For the application described in [18] , the linear support vector machine algorithm as provided by *LIBLINEAR* [6] is used. However, with this approach, a valid value range for the weights cannot be enforced. Specifically, weights can become negative. To reduce the chance of negative weights, artificial training examples are added that require positive weights (setting a single x_i to 1 at a time and the others to 0). These constraints may still be violated in favor of a larger margin or in case of general constraint inconsistencies.

4.4 Dealing with Inconsistent Constraint Sets

Sometimes the set of constraints to be used for learning may be inconsistent because there are constraints that contradict each other. Reasons for this may be manifold – e. g., a user may have changed his mind or the constraints may be from different users or contexts in general. In such case, it is impossible to learn a facet weighting that satisfies all constraints – irrespective of the learning algorithm or the facets used. In order to obtain a consistent set of constraints, the constraint filtering approach described in [12] can be applied as follows:

1. A directed multigraph (i. e., a graph that may have multiple directed edges between two nodes) is constructed with pairs of objects as nodes and the distance constraints expressed by directed edges. For instance, for the distance constraint $d(b, c) < d(a, c)$, a directed edge from the node (b, c) to the node (a, c) would be inserted.
2. All cycles of length 2 are removed, i. e., all directly contradicting constraints. (This can be done very efficiently by checking the graph's adjacency matrix.)
3. The resulting multigraph is further reduced to a Directed Acyclic Graph (DAG) in a randomized fashion: Starting with an empty DAG, the edges of the multigraph are added in random order omitting those edges that would create cycles.
4. The corresponding distance constraints of the remaining edges in the DAG form a consistent constraints set.

This can be repeated multiple times as the resulting consistent set of constraints may not be maximal because of the randomized greedy approach taken in step 3. However, finding a maximum acyclic subgraph would be NP-hard.

4.5   Quadratic Programming Approaches with Soft Constraints

The underlying algorithm [7] of the QP solver used for the approach outlined in Section 4.2 solves convex QP problems of the form

$$\min_{\mathbf{x} \in \Re^n} \mathbf{a}^T \mathbf{x} + \frac{1}{2} \mathbf{x}^T G \mathbf{x} \tag{12}$$

subject to linear equality and inequality constraints

$$\mathbf{x}^T C_e = \mathbf{b}_e \tag{13}$$
$$\mathbf{x}^T C_i \geq \mathbf{b}_i \tag{14}$$

given the vectors $\mathbf{b}_e$ of dimension m_e, $\mathbf{b}_i$ of dimension m_i, and $\mathbf{a}$ of dimension n, and the matrices G of of dimension $n \times n$, C_e of dimension $m_e \times n$, and C_i of dimension $m_i \times n$. The matrix G has to be symmetric positive definite. In this case, a unique $\mathbf{x}$ solves the problem or the constraints are inconsistent. In the original modeling (Section 4.2), the objective is to minimize the weight change under some constraints given the previous weights $w_1^{(old)}, \ldots, w_l^{(old)}$. In particular, this approach can be used to determine facet weights that are closest to a uniform weighting and feasible under the given constraints by simply setting all previous weights to 1. Here, the elements of the vector $\mathbf{x}$ in the QP problem description correspond to the facet weights $w_1, \ldots, w_l$ (where l is the number of facets), and therefore n equals l. The objective function given in Equation 11 can be transformed into:

$$\min_{(w_1,\dots,w_l)\in\Re^l} \sum_{i=1}^{l} w_i^2 - 2 \sum_{i=1}^{l} w_i w_i^{(old)} + \sum_{i=1}^{l} w_i^{(old)\,2} \qquad (15)$$

With respect to Equation 12, the first sum is captured by $\frac{1}{2}\mathbf{x}^T G\mathbf{x}$, the second sum is expressed by $\mathbf{a}^T\mathbf{x}$, and the third sum results in a constant value independent of the w_i and thus can be neglected. A single equality constraint is required to model the bound on the weight sum (Equation 5) by setting the respective coefficients in C_e to 1 and the value in $\mathbf{b}_e$ to l. Setting the i-th element of a row vector of C_i to 1 and the other elements and the corresponding value in $\mathbf{b_i}$ to 0 enforces the non-negativity of w_i. Thus, l inequality constraints are needed to express Equation 4. Finally, each distance constraint is represented by a single row vector of C_i where the value $c_{i,j}$ is the x_i from Equation 7 for the j-th distance constraint. The respective value in $\mathbf{b}_i$ has to be a value $\epsilon > 0$ because 0 would also allow the equality in Equation 6. Naturally, this value should be as small as possible w.r.t. machine precision. Greater values increase the stability of the solution but may also result in an inconsistent system if the solution space is trimmed too rigorously such that no feasible solution exists anymore.

In order to allow a distance constraint as formulated in Equation 7 to be violated, a slack variable $\xi \geq 0$ needs to be introduced such that:

$$\sum_{i=1}^{l} w_i(\delta_{f_i}(s,b) - \delta_{f_i}(s,a)) + \xi > 0 \qquad (16)$$

This has to be done individually for all k distance constraints of the QP problem resulting in the respective slack variables $\xi_1,\dots,\xi_k$. They are modeled as k additional dimensions of the vector $\mathbf{x}$ which is now $(w_1,\dots,w_l,\xi_1,\dots,\xi_k)$. (Consequently, the dimensionality of the modified QP problem is $l + k$ and as the number of distance constraints k can become quite big, this has a significant impact on the performance of the optimization algorithm.) For the constraints that ensure the weight bounds and the non-negativity of the weights, the added matrix columns are filled with zeros. For each of the k distance constraints, only the value in the column of the respective slack dimension is set to 1 while all others remain zero. Slack values other than 0 have to result in a penalty. To this end, the objective function needs to be extended. There are two possibilities to incorporate a slack penalty: either in the linear or the quadratic part of Equation 12. In the first case, the sum of the slack values is minimized, whereas in the second case, it is the sum of the squared slack values. In order to allow a balance between the (initial) objective function and the minimization of the aggregated slack, the latter is weighted by a constant κ. In case the sum of the squared slack values is used, both, negative and positive slack values are penalized. This way, the solver naturally avoids negative slack values. For the (linear) sum, however, the QP approach does not work if negative slack values are allowed because this introduces the problem that a single big negative slack value (of a constraint which is not violated) can compensate many small positive slack values of constraints that are violated. This results in a bias towards solutions with more violated constraints than necessary. Therefore, k inequality constraints that explicitly demand non-negative slack values have to be added to the scheme. Both approaches for incorporating a slack penalty into the objective function of the QP solver can be combined with the original primary objectives of minimizing the weight change. Furthermore, it is possible to have no primary objective and just minimize the slack penalty. The performance of these combinations is analyzed in the following section.

5   Experimental Comparison

In order to compare the different adaptation approaches covered above in a fully controlled environment, an experiment has been conducted using the publicly available *Magnatagatune* benchmark dataset [9]. The setup is explained in the following including the dataset and its pre-processing (Section 5.1), the constraint sets and adaptation algorithms (Section 5.2), and the evaluation methodology (Section 5.3). Results are presented and discussed in Section 5.4.

5.1   Dataset and Pre-processing

The *Magnatagatune* dataset comprises 25863 clips – each one 29 seconds long – generated from 5405 source MP3s provided by the American independent record label *Magnatune* for research purposes. The clips are annotated with a combination of 188 unique tags that have been collected through the *TagATune* game [9]. Additionally, the dataset contains a detailed analysis of each clip computed using the *EchoNest* API.[2] The features comprise musical events, beats, structure, harmony, and various global attributes such as key, mode, loudness, tempo and time signature. Most importantly for the purpose of this evaluation, there is also a set of music similarity judgments. This information has been collected by showing a triple of clips and asking the player to choose the most different one. 533 such triples have been presented to multiple players resulting in 7650 similarity judgments.

In total, 110 facets are used to describe the distances between the clips in the experiment. An overview with brief explanations is given in Table 1. The facets comprise seven globally extracted features of which two – dancability and energy – are not contained in the original clip analysis information of the dataset but have become available with a newer version of the *EchoNest* API. Furthermore, the segment-based features describing pitch ("chroma") and timbre have been aggregated (per dimension) resulting in 12-dimensional vectors with the mean and standard deviation values. This has been done according to the approach described in [5] for the same dataset. The 188 unique tags used in the manual annotations have been preprocessed as follows:

---

[2] A detailed description of the extracted features can be found in the documentation of the *EchoNest Track Analyze* API under http://developer.echonest.com/

Table 1. Facet definition for the *Magnatagatune* dataset used in the experiment. Top rows: Globally extracted features. Middle rows: Aggregation of features extracted per segment. Bottom row: Manual annotations from *TagATune* game.

feature	dim	value description	distance measure
key	1	0 to 11 (one of the 12 keys) or -1 (none)	binary (exact match)
mode	1	0 (minor), 1 (major) or -1 (none)	binary (exact match)
loudness	1	overall value in decibel (dB)	absolute difference
tempo	1	in beats per minute (bpm)	absolute difference (& tempo doubling)
time signature	1	3 to 7 ($\frac{3}{4}$ to $\frac{7}{4}$), 1 (complex), or -1 (none)	binary; $\delta(3,6) = 0.5$
danceability	1	between 0 (low) and 1 (high)	absolute difference
energy	1	between 0 (low) and 1 (high)	absolute difference
pitch mean	12	dimensions correspond to pitch classes	Euclidean distance
pitch std. dev.	12	dimensions correspond to pitch classes	Euclidean distance
timbre mean	12	normalized timbre PCA coefficients	Euclidean distance
timbre std. dev.	12	normalized timbre PCA coefficients	Euclidean distance
tags (99 facets)	1	binary, one facet per tag, very sparse	binary (exact match)

1. Merging of singular and plural forms (e. g., "guitar" and "guitars").
2. Spelling correction (e. g., "harpsicord" → "harpsichord").
3. Combination of semantically identical tags (e. g., "funk" and "funky").
4. Creation of meta-tags with higher coverage for groups of tags that express the same concept. (e. g., "instrumental" = "instrumental" or "no vocal(s)" or "no voice(s)" or "no singer(s)" or "no singing").
5. Removal of unused tags (w.r.t. the relevant subset of *Magnatagatune*).

The resulting 99 tags are interpreted as one (binary) facet each.[3]

5.2   Constraint Sets and Algorithms

Two distance constraints can be derived from a single judgments that clip c is the most different of a triple (a, b, c), namely $d(a, b) < d(a, c)$ and $d(a, b) < d(b, c)$. However, the resulting set of constraints is inconsistent because there are constraints that contradict each other. This is most likely because the similarity judgments stem from multiple players of the *TagATune* game. Applying the filtering technique described in Section 4.4, a constraint graph with 15300 edges of which 1598 are unique is constructed. After the removal of length 2 cycles, 860 unique edges remain (6898 in total). The randomized filtering finally results in a DAG with 674 unique edges (6007 in total). Thus, the filtered consistent set

---

[3] Alternatively, it is possible to combine all annotations into a single facet or define facets for groups of related tags (e. g., all tags related to instrumentation) which significantly reduces the number of facets. However, this would also drastically reduce the size of the *selected constraints* set described in the following.

contains 674 constraints of which each is backed by 8.9 judgments on average. In the following, this set is referred to as *all constraints* set.

Even for the consistent *all constraints* set, it is impossible to learn a facet weighting that violates none of the constraints because the information captured by the facets is insufficient. I. e., players may have considered aspects in their judgments that are not covered by the features. From the classification perspective of Equation 7 this means that there is no hyperplane that clearly separates the positive from the negative examples. Or from the QP perspective, the system of equality and inequality constraints is inconsistent – i. e., it has no solution. Therefore, another set – in the following referred to as *selected constraints* – has been constructed by further filtering the *all constraints* set. To this end, the randomized approach of Section 4.4, step 3 has been applied again but this time constraints are only added to the set if the resulting QP problem has a solution. The *selected constraints* set obtained this way contains 521 constraints.

At first sight, the two sets of constraints seem to be quite large. After all, which user would like to answer several hundred questions of the form *"Which one of these three objects do you think is the most distinct one from the others?"* However, these distance constraints are in fact only the very atomic pieces of information used to guide the adaptation. As the example applications described in earlier work show, usually multiple such distance constraints are derived from a single action like moving an object [17], correcting a ranking [2] or adding a tag annotation [18].

Table 2 lists the considered algorithms and their parameters. They comprise the three algorithms used in the applications described in previous work [17,2,18] and furthermore several variants of the QP approach with added slack dimensions that allow the violation of distance constraints (Section 4.5). As the Grad-Desc learner may get stuck in a local optimum, the computation is repeated up to 50 times if no solution could be found that satisfies all training constraints. Each run uses a different random order of the same training constraints. Finally, the solution which results in the lowest number of constraint violations is chosen.

5.3 Evaluation Methodology

The evaluation aims to answer the following questions:

- How good is the obtained adaptation (in terms of constraint violations)?
- How fast (with how much user effort) can it be learned?
- How stable is the quality of an adaptation if new constraints are added?

The number of violated distance constraints serves as a measure for how well the algorithm has adapted to the similarity preferences given some training constraints. All algorithms except QPmin(Δw) that cannot deal with inconsistencies are tested on both sets of constraints described in Section 5.2. For the *selected constraints* set, a solution satisfying all constraints is expected. Whereas, for the *all constraints* set, the behavior of the algorithms under constraints that cannot all be satisfied is tested. As the size difference between the two sets is 153,

Table 2. Algorithms covered in the comparison. Top: Algorithms used in applications described in earlier work (Sections 4.1 to 4.3). Bottom: Alternative QP problem formulations with added slack dimensions (Section 4.5).

abbreviation	algorithm	parameters
GradDesc	Gradient Descent	50 repeats with random permutations of training samples, dyn. learning rate
QPmin(Δw)	Quadratic Programming	minimal weight change, no slack
LibLinear	Maximal Margin Classifier (Java *LIBLINEAR* v1.5)	L2-regularized L2-loss SVC, $C = 10^7$, $\epsilon = 10^{-6}$, no bias term
QPmin(ξ)	Quadratic Programming	no primary objective, lin. slack $\kappa = 1$
QPmin(ξ^2)	Quadratic Programming	no primary objective, quad. slack $\kappa = 1$
QPmin(Δw+ξ)	Quadratic Programming	min. weight change, lin. slack $\kappa = 1$
QPmin(Δw+ξ^2)	Quadratic Programming	min. weight change, quad. slack $\kappa = 1$
QPmin(Δw+$10^5\xi^2$)	Quadratic Programming	min. weight change, quad. slack $\kappa = 10^5$

the optimal performance value for the *all constraints* set is expected to be close to 150 in terms of constraint violations. For each of the two sets, 100 random permutations of the distance constraints are generated. Each permutation is presented to the adaptation algorithm – one constraint at a time (i. e., stepwise) – until all constraints are used for training. After each step, the number of violated constraints in the whole set is determined. The values are averaged per step over the 100 permutations to reduce ordering effects.

5.4   Results

Figure 1 shows the detailed performance of all algorithms listed in Table 2. Each diagram combines the results of a single algorithm on both constraint sets – *all constraints* (top, red curve) and *selected constraints* (bottom, blue curve) – as these do not overlap. Both colored curves refer to the average of the 100 runs. Additionally, all performance values obtained for the 100 random permutations of the constraints are shown as points (in light gray). This gives an impression of the variance between the different runs. The two gray dotted horizontal lines indicate the baseline performance value obtained by the uniform facet weighting on all constraints (upper line) and the subset of selected constraints (lower line). The scaling of all plots is identical for better comparability. It is also the same for both axis.

Comparing the algorithms applied in earlier work (top rows of Table 2), the plots for the *selected constraints* set, where a weighting can be found that satisfies all constraints, are almost identical. For LibLinear the mean curve is a bit steeper, indicating slightly better early solutions. However, a little more variance can be observed – especially between 50 and 80 training constraints. Furthermore, it has to be noted that GradDesc and LibLinear converge on a solution that leaves a small number of constraints violated whereas QPmin(Δw) finds a weighting without constraint violations. The GradDesc learner still gets stuck in a local optimum

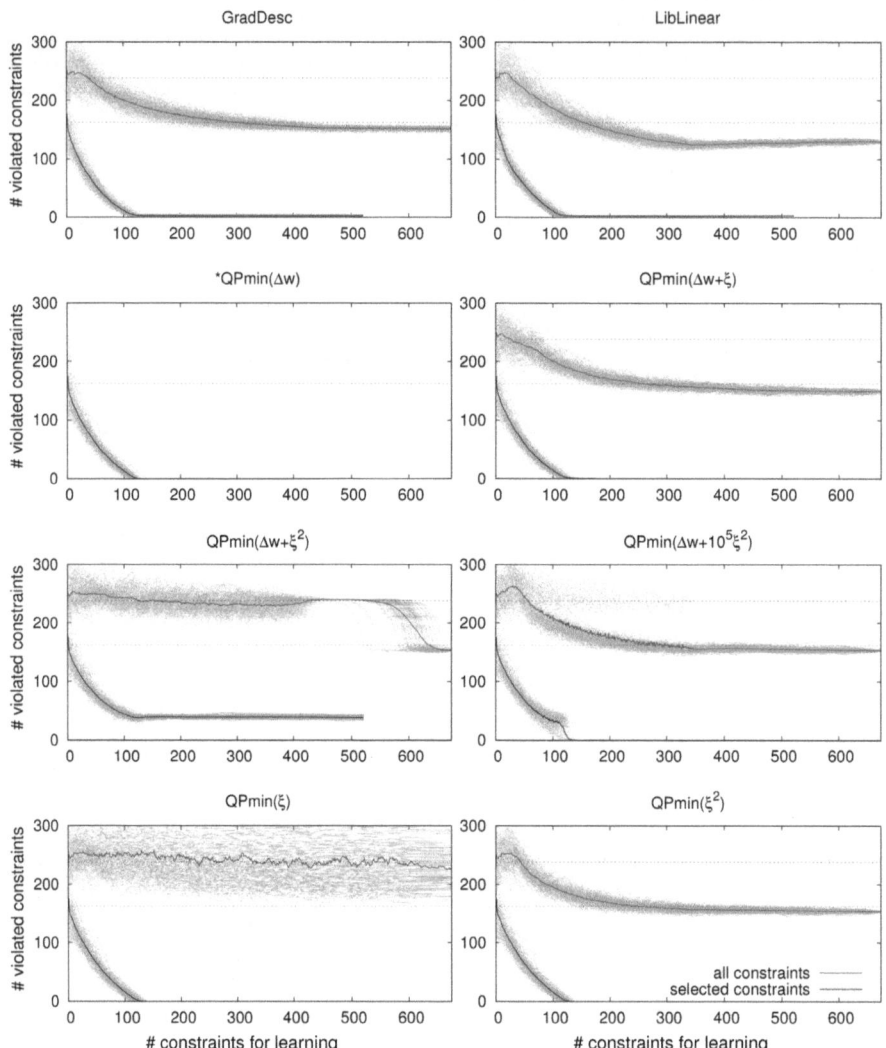

Fig. 1. Performance plots for the algorithms listed in Table 2 averaged over 100 random permutations of the *all constraints* set (red curves) and *selected constraints* set (blue curves). Baseline values for uniform facet weightings are shown as dotted gray horizontal lines. The gray point clouds visualize the value distributions within the 100 runs. (*QPmin(Δw) is not applicable on the *all constraints* set.)

possibly close to the global one. The problem of LibLinear is that it favors a large margin over small constraint violations, e. g., caused by small negative facet weights. For the larger *all constraints* set, QPmin(Δw) does not return a solution because the derived QP system to be solved is inconsistent. Comparing GradDesc and LibLinear which both can deal with constraint violations, the latter again shows faster convergence but slightly higher variance. Most notably, LibLinear leads to a solution that violates approximately 30 constraints less than GradDesc. The main reason for this is that it trades a few (slightly) violated weight bounds constraints against a larger number of distance constraints that are not violated. This results however in an invalid weighting.

Looking at the alternative QP approaches that aim to minimize only the slack without a primary objective, the plots for the *selected constraints* set look almost identical. They do not differ much from QPmin(Δw) even though the objective is much different here. Therefore, it can be concluded that both approaches work well if there is a solution that violates no constraints. However, the performance could not look much more different for the *all constraints* set: QPmin(ξ), i. e., modeling the slack in the linear part of the QP objective function (and leaving the quadratic part constant), seems not to work at all. There is almost no improvement compared to the baseline. At the same time, the variance increases which is much in contrast to all other approaches. However, QPmin(ξ^2), i. e., modeling the slack in the quadratic part of the objective function, produces a better solution than GradDesc. At the beginning, for up to 30 training constraints, there is no improvement. In fact both plots look here very much alike. But then the number of violations drops quickly and for 100 training constraints, it is already lower than for GradDesc.

For the alternative QP approaches that minimize the change of the facet weights with soft constraints, QPmin(Δw+ξ), has the best performance on both sets of constraints. This is surprising considering that QPmin(ξ) does not work at all for the *all constraints* set. Much in contrast, minimizing the quadratic slack penalty works only well without the primary objective (QPmin(ξ^2)). For the combination, QPmin(Δw+ξ^2), there seems to be a conflict between both objectives. This results in an unsatisfactory adaptation for the *selected constraints* set with more than 40 constraint remaining violated. The QPmin(Δw+ξ^2) plot for the *all constraints* set is very remarkable. It can be divided into three sections: In the first section up to roughly 440 constraints, there is high variance between the permutations and no significant improvement. Then, however, the values converge and until about 525 constraints, no variance can be observed. This point coincides with the size of the *selected constraints* set which is close to the maximal number of constraints that can be satisfied. Afterwards, in the last section, the number of violated constraints quickly decreases to a final value that is comparable to the other working approaches. This late adaptation suggests that the primary objective (minimizing the weight change) suppresses the minimization of the slack until the last section. Indeed, the facet weights have converged to 1 at the beginning of the second section which explains the performance close to the baseline (uniform facet weights). Only afterwards, the importance of the slack gains the upper hand – most likely because of the high number of slack dimensions caused by the many training constraints in this section. Choosing a high slack weight results in an earlier adaptation as shown for QPmin(Δw+$10^5\xi^2$) with $\kappa = 10^5$. However, the variance is very high and the performance is still inferior to QPmin(Δw+ξ). Even higher values of κ result in no significant improvement.

Finally, Table 3 lists some empirically determined values for the processing time of the different algorithms. Especially in interactive settings, a short response time is important. For GradDesc, the times refer only to a single repetition. Generally, these measurements can only give an impression of the processing time for the adaptation as no special preparations of the testing system, the runtime environment or the compiler have been made. The values are averaged over 100 random permutations and have been measured for 10, 100 and all available training constraints of the two sets. Values for the *all constraints* set are expected to be higher because of the unavoidable constraint violations that occur here. For GradDesc and LibLinear, this is surprisingly not the case. Possibly, finding a solution for the *selected constraints* set is harder for these algorithms. However, it has to be noted that LibLinear is the only approach that in the current implementation of the library interface requires slow hard disk access to read the problem description from a temporary file. The actual processing times for LibLinear are therefore much lower. In the adaption experiment described in [18] with much larger constraint sets, LibLinear has already shown that its runtime scales well. The QP approaches could run into problems here – especially QPmin(Δw+ξ) which requires even more constraints for the non-negativity of the slack variables. In contrast to the other algorithms, GradDesc, which is rather slow (but also the only algorithm in the evaluation that does not rely on highly optimized library code) could be interrupted during computation and still return a satisfying adaptation.

A direct comparison of all tested approaches is shown in Figure 2. For the *selected constraints* set, all approaches except those using the square slack penalty work almost equally well. I. e., if a solution exists that satisfies all constraints, one is found. Only GradDesc gets stuck a little too early and LibLinear favors the larger margin. Of the approaches with square slack penalty, QPmin(Δw+ξ^2) does not work well leaving roughly 40 unsatisfied constraints and QPmin(Δw+$10^5\xi^2$) converges too slowly. For the harder *all constraints* set, LibLinear can be considered as the "disqualified winner" of the competition. It shows the overall quickest convergence requiring less steps then the other approaches for the adaptation and returns weightings that violate significantly fewer constraints. However, the latter is only possible because of "cheating" as the weights violate the essential non-negativity constraint (Equation 4) and thus cannot be interpreted as intended. Given these results and the good scalability for large problems, an internal modification of *LIBLINEAR* that ensures non-negative weights looks promising. However, this is not a trivial task and intended for future work. QPmin(Δw+ξ) has the best final performance value for a valid adaptation on *all constraints* which is even slightly below 150. However, its adaptation is a bit slow

Table 3. Processing times (in seconds) for the adaptation depending on the number of training constraints measured on both constraint sets. Values averaged over 100 repetitions on a consumer notebook (2.4 GHz Intel Core 2 Duo, 4GB RAM). Algorithms without satisfying adaptations in the evaluation are in gray.

algorithm (abbreviation)	selected constraints			all constraints		
	10	100	521	10	100	674
GradDesc	<0.01	3.34	13.29	<0.01	5.41	8.45
QPmin(Δw)	<0.01	0.02	1.08			
LibLinear	0.13	1.21	2.54	0.38	0.46	1.19
QPmin(ξ)	<0.01	0.06	5.84	<0.01	0.19	19.62
QPmin(ξ^2)	<0.01	0.03	1.15	<0.01	0.05	5.95
QPmin($\Delta w+\xi$)	<0.01	0.06	6.59	<0.01	0.06	27.73
QPmin($\Delta w+\xi^2$)	<0.01	0.02	0.82	<0.01	0.03	3.05
QPmin($\Delta w+10^5\xi^2$)	<0.01	0.02	1.12	<0.01	0.04	4.59

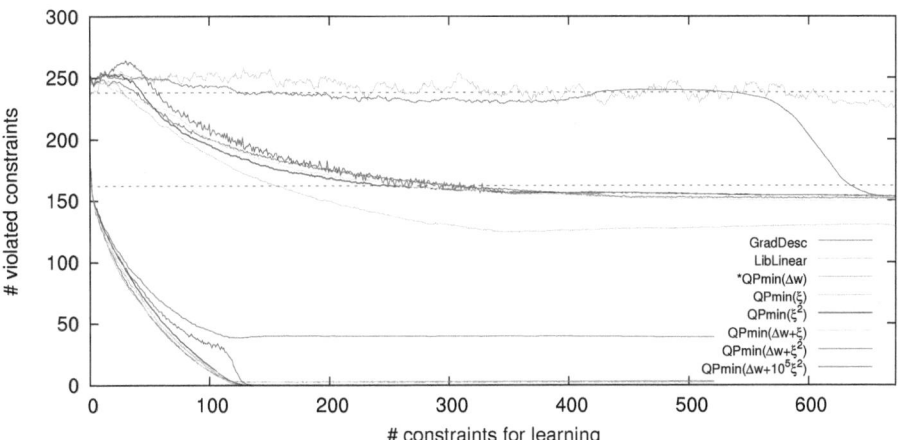

Fig. 2. Direct comparison of all approaches tested in the experiment. All values are averaged over 100 random permutations of the constraints. (*QPmin(Δw) is not applicable on *all constraints* set.)

in the beginning. For the first 70 steps, GradDesc would be a better choice and in the middle section, QPmin(ξ^2) does slightly better. In the end, the performance difference of these three approaches is only very small. Finally, QPmin(ξ) does not work at all for *all constraints* and QPmin($\Delta w+\xi^2$) converges only in the end which is not acceptable either. These combination should therefore not be used.

6   Conclusions

Based on a generalized view that brings together our work from recent years in the field of adaptive similarity and distance measures based on weighted linear combinations, we have conducted an experimental comparison of the different approaches. To this end, we have utilized the publicly available *Magnatagatune* benchmark dataset. Pre-processing of this dataset comprised the refinement of the tag annotations, the definition of facets, the generation of distance constraints from the similarity judgments, and the filtering of the constraints to obtain a consistent set. As performance measure, the number of constraint violations has been plotted against the number of constraints used for training. In general, the evaluation methodology is not confined to approaches that model distance as a weighted linear combination of facets. Basically, any algorithm that uses relative distance constraints can be tested this way. Thus, it is possible to extend the comparison and also cover approaches that, e. g., use the Mahalanobis distance or complex kernel-based models which is planned for future work.[4] Further plans include to modify *LIBLINEAR* such that weights cannot become negative.

Acknowledgments. This work was supported in part by the German National Merit Foundation and the German Research Foundation (DFG) under the project AUCOMA.

References

1. Bade, K.: Personalized Hierarchical Structuring. PhD thesis, Otto-von-Guericke-University Magdeburg (2009)
2. Bade, K., Garbers, J., Stober, S., Wiering, F., Nürnberger, A.: Supporting folk-song research by automatic metric learning and ranking. In: Proc. of the 10th Int. Conf. on Music Information Retrieval (ISMIR 2009) (2009)
3. Bade, K., Nürnberger, A.: Creating a cluster hierarchy under constraints of a partially known hierarchy. In: Proc. of the 2008 SIAM Int. Conf. on Data Mining (2008)
4. Cheng, W., Hüllermeier, E.: Learning similarity functions from qualitative feedback. In: Althoff, K.-D., Bergmann, R., Minor, M., Hanft, A. (eds.) ECCBR 2008. LNCS (LNAI), vol. 5239, pp. 120–134. Springer, Heidelberg (2008)
5. Donaldson, J., Lamere, P.: Using visualizations for music discovery. In: Tutorial at the 10th Int. Conf. on Music Information Retrieval (ISMIR 2009) (2009)
6. Fan, R., Chang, K., Hsieh, C., Wang, X., Lin, C.: Liblinear: A library for large linear classification. The Journal of Machine Learning Research 9, 1871–1874 (2008)
7. Goldfarb, D., Idnani, A.: A numerically stable dual method for solving strictly convex quadratic programs. Mathematical Programming 27(1), 1–33 (1983)
8. Joachims, T.: A Support Vector Method for Multivariate Performance Measures. In: Proc. of the Int. Conf. on Machine Learning (ICML 2005) (2005)

---

[4] The constraint sets derived from the *Magnatagatune* dataset can be provided upon request. Please contact `sebastian.stober@ovgu.de`

9. Law, E., von Ahn, L.: Input-agreement: a new mechanism for collecting data using human computation games. In: Proc. of the 27th Int. Conf. on Human Factors in Computing Systems (CHI 2009) (2009)
10. Lübbers, D., Jarke, M.: Adaptive multimodal exploration of music collections. In: Proc. of the 10th Int. Conf. on Music Information Retrieval (ISMIR 2009) (2009)
11. McFee, B., Barrington, L., Lanckriet, G.: Learning similarity from collaborative filters. In: Proc. of the 11th Int. Conf. on Music Information Retrieval (ISMIR 2010) (2010)
12. McFee, B., Lanckriet, G.: Heterogeneous embedding for subjective artist similarity. In: Proc. of the 10th Int. Conf. on Music Information Retrieval (ISMIR 2009) (2009)
13. Nürnberger, A., Klose, A.: Improving clustering and visualization of multimedia data using interactive user feedback. In: Proc. of the 9th Int. Conf. on Information Processing and Management of Uncertainty in Knowledge-Based Systems (IPMU 2002) (2002)
14. Salton, G., Buckley, C.: Term weighting approaches in automatic text retrieval. Information Processing & Management 24(5), 513–523 (1988)
15. Slaney, M., Weinberger, K.Q., White, W.: Learning a metric for music similarity. In: Proc. of the 9th Int. Conf. on Music Information Retrieval (ISMIR 2008) (2008)
16. Nürnberger, A., Stober, S.: User modelling for interactive user-adaptive collection structuring. In: Boujemaa, N., Detyniecki, M., Nürnberger, A. (eds.) AMR 2007. LNCS, vol. 4918, pp. 95–108. Springer, Heidelberg (2008)
17. Stober, S., Nürnberger, A.: Towards user-adaptive structuring and organization of music collections. In: Detyniecki, M., Leiner, U., Nürnberger, A. (eds.) AMR 2008. LNCS, vol. 5811, pp. 53–65. Springer, Heidelberg (2010)
18. Stober, S., Nürnberger, A.: Similarity adaptation in an exploratory retrieval scenario. In: Detyniecki, M., Knees, P., Nürnberger, A., Schedl, M., Stober, S. (eds.) AMR 2010. LNCS, vol. 6817, pp. 144–158. Springer, Heidelberg (2012)
19. Stober, S.: Adaptive distance measures for exploration and structuring of music collections. In: Proc. of AES 42nd Conf. on Semantic Audio (2011)
20. Wolff, D., Weyde, T.: Combining Sources of Description for Approximating Music Similarity Ratings. In: Detyniecki, M., García-Serrano, A., Nürnberger, A., Stober, S. (eds.) AMR 2011. LNCS, vol. 7836, pp. 114–124. Springer, Heidelberg (2013)

Combining Sources of Description
for Approximating Music Similarity Ratings

Daniel Wolff and Tillman Weyde

City University London, Department of Computing
Northampton Square, London EC1V 0HB, UK
daniel.wolff.1@soi.city.ac.uk,
t.e.weyde@city.ac.uk

Abstract. In this paper, we compare the effectiveness of basic acoustic features and genre annotations when adapting a music similarity model to user ratings. We use the Metric Learning to Rank algorithm to learn a Mahalanobis metric from comparative similarity ratings in in the MagnaTagATune database. Using common formats for feature data, our approach can easily be transferred to other existing databases. Our results show that genre data allow more effective learning of a metric than simple audio features, but a combination of both feature sets clearly outperforms either individual set.

Keywords: Music Information Retrieval, Music Recommendation, Computational Modelling, Music Similarity, Music Perception.

1 Introduction

Adapting music recommendation systems to the needs or preferences of users is a critical factor in the success of commercial music sites today. Presenting relevant results to users promises to increase the aesthetical, social or financial benefit of recommendation. Depending on the context and intention, different ways of determining relevance in music may be appropriate to fulfill the expectations of users and businesses.

Our focus lies on generating models of perceived or stated music similarity for acoustic recordings of music, which can be applied in music exploration or recommendations systems. To this end, we exploit user ratings from a human computation source, which yield relative similarity ratings about triples of songs. The raw data is approximated using binary rankings expressing "Songs {A, ...} are more similar to Song B than the Songs {C, ...}". Such rankings are used for constraining the optimisation of metrics, which are defined on the vector space of features describing the music. There are several algorithms available for this task. In this paper, we choose the MLR algorithm for its robust behaviour, and focus on the effects of using content-based features compared to genre annotations as well as different representations of these for learning the metrics.

M. Detyniecki et al. (Eds.): AMR 2011, LNCS 7836, pp. 114–124, 2013.

1.1   Related Work

The selection of features suitable for a specific task has been a field of active research, relevant to many disciplines within Music Information Retrieval. Properties of features and their selection routines, besides customisation to users, allow for a definition and selection of relevant information, the structuring of datasets and thus for specialised indexing methods to be used.

In 2001, Pickens categorised selection techniques for music information retrieval on symbolic data, focusing in the relation of musicological properties of the extraction routines and their implications for retrieval performance [1]. Novello et al. performed a study on music similarity perception [2]. Asking subjects to select pairs of best- and worst fitting songs out of triplets presented, they rendered a musical similarity space using multidimensional scaling. Their evaluation of the data gives important insights in the concordance within similarity ratings of musicians and non-musicians, and the correlation of these with musical genres.

In the evaluation and optimization chapter of his dissertation [3] Pampalk extensively evaluates the performance of 14 content-based features in a genre-classification task: The correlation of songs' genres and clusters inferred from a similarity defined by weighted feature influences are compared, using leave-one-out cross-validation. The tests were performed on several databases with sizes of 100 to 15335 western pop and classical music tracks. Moreover, combinations of the six best-performing features were evaluated using a combinatorial approach. Results showed that spectral features have a strong weight in best performing configurations, alongside with percussivity and fluctuation pattern features.

A set-based method for learning a feature weighting using an interactive playlist-based user survey has been presented by Allan et al. [4]. Users could specify their preferred similarity concepts using two example song sets, one for similar and one for dissimilar songs. Moreover, recommendations were improved using a feedback loop.

Barrington et al. [5] used timbral and harmonic features, as well as tags and information mined from the web for text-based audio retrieval. Different ways of combining these information sources – calibrated score averaging, RankBoost, and kernel combination support vector machines – were evaluated. As shown in [6], the kernel combination approach enabled a straightforward analysis of the different features' influences. McFee et al. [7] have designed an algorithm for learning a Mahalanobis metric to rankings, based on the Cutting-Plane Structural SVM training algorithm of [8]. They used it in a hybrid approach for parametrising a purely content-based music similarity metric using collaborative filtering data. Their content-based classifiers were successfully applied for music discovery in the so-called long tail, i.e. sets of sparsely annotated and barely documented music, e.g. new or less popular songs, where this method enables improved recommendation.

Other algorithms for learning Mahalanobis metrics from comparative user ratings have been published: Schultz and Joachims [9] trained a weighted Euclidean distance metric using relative comparisons. In [10], Davis et al. formulate a metric learning problem similar to the above, as an LogDet-optimisation task. Their approach uses another arbitrarily predefined Mahalanobis metric for the regularisation target.

A comparison of several algorithms for similarity adaptation, using similar data based on MagnaTagATune has been presented at the AMR11 conference by Stober et al. [11]. Their approach and results will be related to ours in Section 5.1.

2   Music and Descriptions

The majority of the data used for the experiments in Section 4 is based on the MagnaTagATune dataset. TagATune is a web-based[1] game, collecting tags associated with certain songs in a human-computation manner. Furthermore, in a bonus mode of the game, user votes on perceptual outliers out of song triplets are collected: Users have to agree on a song out of three which is the least similar to the remaining songs. Documenting the application of this game on a song database from the Magnatune label, MagnaTagATune combines the audio content, derived features and tagging information of 25863 30-second audio clips into a publicly available dataset [12]. The data from the bonus mode contains 7650 individual votes on 533 triplets of clips. The clips C_i, $i \in \{1, \cdots 1019\}$ included in these triplets constitute the dataset used in our experiments.

2.1   Genre Annotations

We extend the information in this dataset by extracting the genre tags the Magnatune label assigned to the clips' corresponding albums for indexing and marketing purposes. This information is publicly available via their xml catalogue[2]. The catalogue contains ordered genre descriptions which exhibit a hierarchical character, which is ignored in this application. Each clip in our experiment subset is tagged with around 2-4 genre descriptions. Thereby, a vocabulary of 44 genre tags is established. The genre information for an individual clip C_i is now expressed using binary feature vectors $F_i^{genre} \in \{0,1\}^{44}$, each component corresponding to whether the clip is annotated with a particular genre description.

2.2   Content-Based Features

The content-based features contained in the MagnaTagATune dataset have been created using the "The Echo Nest" API 1.0. The algorithms used in the API have been described in [13]. We use the segment-based chroma and timbre information for each clip to generate a single feature vector describing the entire clip.

---

[1] http://www.gwap.com/gwap/gamesPreview/tagatune/
[2] http://magnatune.com/info/api.html

The features used here are intended to represent the rough harmonic and timbral content of each clip. This is achieved by separately clustering the chroma and timbre vectors into four clusters $t_i^j \in \mathbb{R}^{12}$, $c_i^j \in \mathbb{R}_{\geq 0}^{12}$, $j \in \{1, \cdots 4\}$ for each clip C_i, $i \in \{1, \cdots 1019\}$. As the temporal segments related to each of the chroma and timbre vectors are of different length, we use a weighted k-means variant, including the single feature vectors' corresponding segment lengths for determining the cluster centroids. Vectors only accounting for a short frame of time thus have less impact in determining one of the cluster centroids. The resulting centroids are sorted in descending order, using the accumulated weights of the corresponding segments. These are represented in the scalars $\lambda(c_i^j), \lambda(t_i^j) \in [0, 1]$. The chroma centroids are then normalised using

$$\tilde{c}_i^j = \frac{c_i^j}{\max_k(c_i^j(k))}. \tag{1}$$

As the components of the timbre centroids feature strong outliers, these were clipped to the percentile $p_t^{0.85}$ corresponding to the interval $\left[0, p_t^{0.85}\right]$ including 85% of the absolute component values $|t_i^j|$ of all clusters j and clips C_i. Afterwards, the clipped values were shifted and scaled to fit the interval $[0, 1] \ni \tilde{t}_i^j$.

Finally, the above values are combined into feature vectors

$$F_i^{audio} = \left(\tilde{c}_i^1, \cdots \tilde{c}_i^4, \lambda(c_i^1) \cdots \lambda(c_i^4), \tilde{t}_i^1, \cdots \tilde{t}_i^4, \lambda(t_i^1) \cdots \lambda(t_i^4)\right)^T \in \mathbb{R}^{104}. \tag{2}$$

2.3 Combined Features

In order to combine the information from genre and audio features, both feature vectors are concatenated into the combined feature vector

$$F_i^{comb} = \left(F_i^{audio}(1), \cdots F_i^{audio}(104), F_i^{genre}(1), \cdots F_i^{genre}(44)\right)^T \in \mathbb{R}^{148}. \tag{3}$$

2.4 PCA, Reduced Features

For each of the single and combined features, a Principal Component Analysis (PCA) is performed. After sorting according to variance in the principal components, we reduce the dimensionality of the transformed features, keeping only the 20 components with greatest variance across the 1019 clip dataset. In this way we gain a set of three different features sharing a constant dimensionality.

After transformation into the principal component space, across the whole dataset, the individual feature components are shifted and scaled to fit the interval of $[0, 1]$. The impact of such normalising of the features was tested. we found that normalising the features after transformation generally improved the results of the following metric learning. In our experiments below, we call these PCA-reduced and normalised features $\tilde{F}_i^{genre}, \tilde{F}_i^{audio}, \tilde{F}_i^{comb} \in \mathbb{R}^{20}$.

2.5   Binary Rankings

As mentioned above, the dataset contains a set of vote statistics $H = \{H_u \mid u \in \{1, \cdots 1019\}\}$ with vote numbers $H_u = \{h_i, h_j, h_k\}$ on an outlying clip given three clips C_i, C_j, C_k. We gather the binary rankings, used as ground truth in our experiments, by approximating these voting statistics: For each such triplet of songs, where possible, a "winning" outlier is determined, by following the majority of the votes. Let $h_i \leq h_j \leq h_k$. Triplets not featuring unequivocal votings are dismissed.

$$\nexists H_u \in H : h_j = h_k. \tag{4}$$

In order to render the data accessible to other algorithms (see our paper in press [14]), further filtering is applied to minimise the amount of inconsistency in between two different triplets $H_u = \{h_i, h_j, h_k\}$ and $H_v = \{h_i, h_j, h_l\}$, $u \neq v$ with data referring to common clips C_i and C_j:

$$\nexists H_u, H_v \in H : \quad (h_i = \max(H_u) \wedge h_j = \max(H_v))$$
$$\vee \, (h_j = \max(H_u) \wedge h_i = \max(H_v)). \tag{5}$$

From the remaining triplets, we generate 533 binary rankings. Each ranking is defined by two sets r_i^s and r_i^d, where r_i^s contains relatively more and r_i^d less similar clips to a given clip C_i. These rankings very roughly approximate the above user votings, and, when compared to other representations of the MagnaTagATune comparison data, show a greater applicability to the metric learning algorithm explained below. Only 12 sets r_i^d and r_i^a contain more than one clip. Thus, most of the rankings can be read as information about individual clips C_i, C_j, C_k, with $C_j, r_i^s = \{j\}$ being more similar to the query clip C_i than $C_k, r_i^d = \{k\}$.

Note that the comparison of binary rankings used in this method, e.g. in the training of the metric or evaluation of a metric's performance, is only based on the relative ranking positions of clips: The correctness of a ranking is defined by evaluating the relative positions of results marked as more or less similar; a correct ranking positions the more similar clips before the less similar ones.

3   Metric Learning to Rank

McFee et al. developed an algorithm for learning a Mahalanobis distance from binary rankings [7]. The Mahalanobis metrics described here resemble a weighted Euclidean metric, but they also allow for a weighting according to rotation and translation of the vectors. Given two vectors $x, y \in \mathbb{R}^N$, the family of Mahalanobis metrics can be expressed as

$$d_W(x, y) = \sqrt{(x - y)^T W(x - y)}, \tag{6}$$

where $W \in \mathbb{R}^{N \times N}$ is a positive semidefinite matrix, parametrising the distance function. Technically, these distance functions also include pseudometrics, which allow for a zero distance between two non-identical vectors.

The distance function is optimised using an algorithm based on Structural SVM [15]. Using a constrained regularisation approach, the matrix W is determined by comparing possible correct and incorrect rankings and their corresponding parametrisation to W. A feature map ψ, combining feature data and rankings, is given by the matrix valued partial order feature, described in [16]. Used in the sense as below, it emphasizes directions in feature space which are correlated with correct rankings. Given a set of training query feature vectors $q \in X \subset \mathbb{R}^N$ and the associated training rankings y_q^*, the complete quadratic optimisation problem is given by

$$\min_{W,\xi} \quad tr(W^T W) + c\frac{1}{n} \sum_{q \in X} \xi_q,$$

$$\text{s.t.} \quad \forall q \in X, \ \forall y \in Y \setminus \{y_q^*\} :$$

$$\langle W, \psi(q, y_q^*) \rangle_F \geq \langle W, \psi(q, y) \rangle_F + \Delta(y_q^*, y) - \xi_q,$$

$$W_{i,j} \geq 0, \xi_q \geq 0. \tag{7}$$

The ξ_q allow for some of the training constraints to be violated. Here, c determines the balance between the regularisation and ranking loss term. $\langle *, * \rangle_F$ denotes the Frobenius matrix product. The ranking-loss term $\Delta(y_q^*, y)$ assures the margin between the given training rankings y_q^* and incorrect rankings y. Common evaluation measures for information retrieval systems are used to determine the respective minimal margin sizes. We selected the AUC-related methods for our experiments, being more robust than nearest neighbour approaches considering the rankings carry sparse information: The AUC curve compares the relation of true positives and false positives in the ranking calculated using the current training state of the metric. Most of the training rankings y_q^* feature just two defined clip positions which either are in correct or incorrect order. As the complete set of possible rankings Y to consider for each training ranking is too large, a cutting-plane approach (see [8]) is used to predict the most violated constraints. The MLR framework is available online[3].

4 Experiments

We applied the MLR algorithm as described in Section 3 for training a distance measure using the rankings from Section 2.5. The experiment described below is part of a series of general experiments on the feasibility of metric learning from user comparisons. Varying the feature types used for the songs' descriptions, the ability of the learned metrics to reproduce the given rankings were compared. In the following experiments, we used the same constraint – regularisation tradeoff factor $c = 10000$ (see Section 3), which was determined to work well in previous experiments using the same ground truth with similar features.

For our experiments we use fivefold cross-validation. The following figures plot the mean performance over the five different partitions for training and test sets.

---

[3] http://cseweb.ucsd.edu/~bmcfee/code/mlr/

For assessing training performance, we measure the percentage of rankings in the test sets to be correctly reproduced by a trained metric. Rankings are fulfilled, if all clips in r_i^s are ranked before any clip in r_i^d.

4.1   Content vs. Annotation

Figure 1 shows the metric learning success curves regarding three different feature types: content-based features only, genre features only and the combination of these features. With a maximal accuracy of 81.8%, the combined approach has a performance strongly exceeding that of the isolated features. The strength of this effect is probably related to the capability of Mahalanobis metrics to model correlations between features of the different types.

Looking at the individual features, the final results of the content-data (73.37%) come very close to the slightly better performing genre features (73.74%). But when considering the smaller training sets, the genre features seem much more effective. The 68.66% baseline using an unweighted Euclidean distance for these features shows that the genre feature space has greater correlation to the users ratings than the content-based features, only allowing for a 61.7% baseline. Relative to the baselines, the performance gain using the MLR training is much greater on the content-based features.

PCA Experiments. The above experiments may very well be influenced by the information density and especially dimensionality of the described feature representations. Therefore we conducted a second experiment, this time using the fixed-dimensional approximations based on Principal Component Analysis (PCA) of the above features. The following experiments were performed with $c = 1000$. Our early experiments underlined that this factor depends on the feature dimension.

When applying the same experiment as above on the PCA features, the resulting learning curves as pictured in Figure 2 closely resemble the situation without PCA (see Figure 1). The baseline Euclidean metric results for both single-medium feature types have dropped less than 2%, but the performances of the trained metrics drop by 5.4% (to 67.92%) for content-based and 2.3% (to 71.48%) for the genre features, showing a significantly lower performance after training.

Here, the metric based on the combined features achieves a performance of 76.9%, indicating an informational gain by combining the two feature types, instead of just adding dimensions to parametrise. Moreover, the combined features' baseline now exceeds the performance of both of the baselines related to the single features.

Generalisation. When considering the performance on the training data, for the raw features, the metric based on genre features performs worse than the one based on content-based features. The content-based features seem less enabling the learning of a general perceptual trend, specifically fitting to the training rankings.

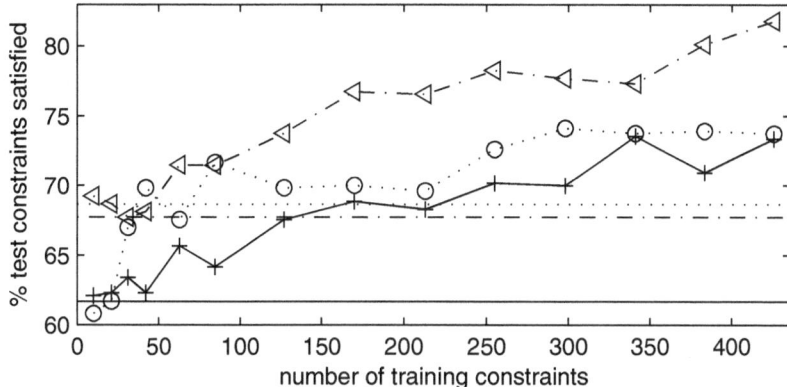

Fig. 1. Results for increasing training set size. Plotted are the mean percentages of fulfilled rankings in the test sets. Top to bottom: Combined features (line-dotted, ◁), genre features (dotted, ○), and content-based features (continuous line, +). The performances of the Euclidean metrics are represented by the straight lines at the bottom, line shapes represent feature types as above.

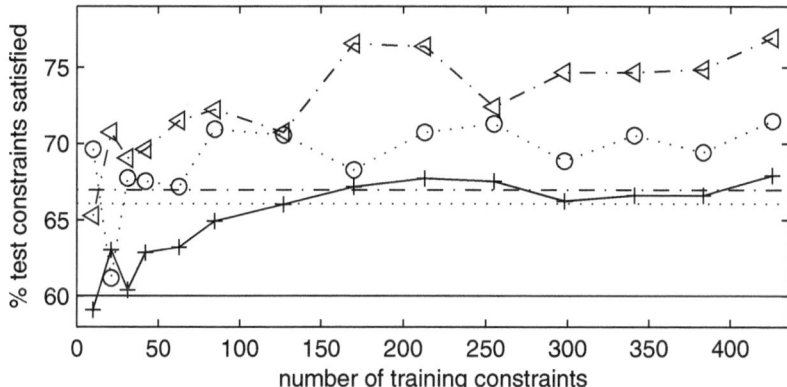

Fig. 2. Results for the PCA reduced features, increasing the training set size. Plotted are the mean percentages of fulfilled rankings. Combined features (line-dotted, ◁), genre features (dotted, ○), and content-based features (continuous line, +). The performances of the Euclidean metric are represented by the straight lines at the bottom.

The results using PCA-reduced features show a strong decrease in the adaptation ability of the metrics. The genre- and content-based features now are almost even in performance on the training data, but in analogy to the above case, the metrics based on genre features perform better on the test sets. This also underlines a stronger correlation of the users ratings and our genre features.

Table 1. Training success for different feature types. Noted are percentages of **training set rankings** correctly reproduced after training with full-size training sets. Statistics are shown for both raw and PCA-reduced feature versions.

	F^{audio}	F^{genre}	F^{comb}
Raw	100 %	92.32%	100%
PCA	85.60%	85.41%	91.28%

5   Discussion

We have used learning of metrics for predicting users' music similarity ratings, comparing the influence of different types of descriptions of the music clips. In line with findings in [3] and others, our experiments show that the combination of content-based and annotated features can improve the adaptability of resulting feature spaces, if they are available. Here, the performance gains achievable with single feature types are well exceeded by the combined feature set. Generally, we observe a strong influence of less data-specific parameters, e.g. dimensionality and number of training examples, on the optimisation process. Further research has to be done towards tuning the procedures for learning metrics to specific properties of the features at hand, like sparseness and dimensionality.

5.1   Comparison with Results in Stober et al. [11]

In these very proceedings, Stober et al. [11] present a comparison of several algorithms for optimising a weighting of several individual similarity measures. Different from our focus on the ability of features to enable a general model of music similarity, the authors cited focus on ability of the algorithms to cope with the user similarity votings, as well as the time efficiency in learning. Therefore, a comparison is given between several procedures for linear optimisation and quadratic programming.

Related to the diagonal-restrained case of MLR, they apply a linear weighting of distance measures based on different features. Most of the content-based features included in the MagnaTagATune dataset are used to represent the clips. For the chroma and timbre information described in Section 2.2, unclustered mean and standard deviation of the segment values are used for aggregation. Moreover, a "danceability feature" is added. In order to include metadata information, Stober et al. process the sparse tag annotations from MagnaTagATune and include it as a 99-dimensional feature. Each of these features has an individual distance function assigned for comparison.

For converting and filtering the similarity information described in Section 2.5, the graph-based approach presented in [17] is used. After filtering inconsistent rankings using a randomized approach, their resulting data contains 674 constraints. Moreover, a second subset of 521 *selected sonstraints* was defined, by evaluating the learnability of the constraints.

In their evaluation, Stober et al. include the training sets in their test sets, focusing on the general ability of the algorithms to adapt to the complete data set. Note that our data shown above shows the mean number of fulfilled constraints for the disjoint test sets of ca. 106 constraints.

For the *selected constraints set*, Stober et al. show several approaches being able to satisfy all (100%) of their constraint data when using a training set size greater than 130. The results on the *all constraints* set show a maximum performance of about 77% of all constraints being fulfilled (we assume 155 violated constraints). GradDesc and other QP approaches reach this maximal adaptation for a training set of about 380 up to 674 constraints.

When including the maximum training set of 426 binary rankings in the calculation of our experimental result, the raw combined feature (F^{comb}) setup (which satisfies 100% of the training set rankings) achieves a mean performance of 96%. This is similar to the results achieved by Stober et al. on their *selected constraints*. In order to allow a meaningful comparison of the results, the datasets have yet to be properly analysed for their individual differences. Especially the *selected constraints* set and the binary rankings used in this study could be analysed for their common information. For their *selected constraints* set, the features and distance measures used by Stober et al. seem to correlate very well with the constraints, allowing for a excellent prediction using only a quarter of the constraints for training.

Assuming our set of binary rankings to be similar, in terms of the contained weak inconsistencies, to the *all constraints* set of Stober et al., our approach appears to be able to fulfill a greater percentage of such constraints. The MLR learning procedure and the feature representation used in this paper allow for a more flexible adaptation: as the Mahalanobis matrix in our feature similarity measure operates directly on the feature data, there are more parameters accessible to the optimisation.

5.2   Future Work

The features used in this study are rather basic in nature, and for the case of the genre features also sparse. More elaborate feature extraction methods may very well improve the performance of the content-based features in particular. Moreover, the representation of the annotations does not accurately reflect the intention of general tag annotations: Usually, the positive information about assigned tags is more important than the information of missing ones. The applied metrics, as linear functions, can not reflect such a bias when using the proposed feature representation.

Further experiments are planned with regards to the representation of the MagnaTagATune triplet comparison data via binary rankings, in order to better represent the individual user votes in the actual training data used for the algorithms.

References

1. Pickens, J.: A survey of feature selection techniques for music information retrieval. In: Proceedings of the 2nd International Symposium on Music Information Retrieval (ISMIR) (2001)

2. Novello, A., Mckinney, M.F., Kohlrausch, A.: Perceptual evaluation of music similarity. In: Proceedings of the 7th International Conference on Music Information Retrieval (ISMIR) (2006)

3. Pampalk, E.: Computational Models of Music Similarity and their Application in Music Information Retrieval. Ph.D. thesis, Vienna University of Technology, Vienna, Austria (2006)

4. Allan, H., Müllensiefen, D., Wiggins, G.: Methodological considerations in studies of musical similarity. In: 8th International Conference on Music Information Retrieval, pp. 473–478 (2007)

5. Turnbull, D.R., Barrington, L., Lanckriet, G., Yazdani, M.: Combining audio content and social context for semantic music discovery. In: SIGIR 2009: Proceedings of the 32nd International ACM SIGIR Conference on Research and Development in Information Retrieval, pp. 387–394. ACM, New York (2009)

6. Barrington, L., Yazdani, M., Turnbull, D., Lanckriet, G.: Combining feature kernels for semantic music retrieval. In: Proceedings of the 9th International Conference on Music Information Retrieval (ISMIR), pp. 614–619 (2008)

7. Mcfee, B., Lanckriet, G.: Metric learning to rank. In: Proceedings of the 27th Annual International Conference on Machine Learning (ICML) (2010)

8. Joachims, T., Finley, T., Yu, C.-N.J.: Cutting-plane training of structural svms. Machine Learning 77, 27–59 (2009)

9. Schultz, M., Joachims, T.: Learning a distance metric from relative comparisons. In: Advances in Neural Information Processing Systems (NIPS). MIT Press (2003)

10. Davis, J.V., Kulis, B., Jain, P., Sra, S., Dhillon, I.S.: Information-theoretic metric learning. In: Proceedings of the 24th International Conference on Machine Learning (ICML 2007), pp. 209–216. ACM, New York (2007)

11. Stober, S., Nürnberger, A.: An experimental comparison of similarity adaptation approaches. In: Detyniecki, M., García-Serrano, A., Nürnberger, A., Stober, S. (eds.) AMR 2011. LNCS, vol. 7836, pp. 96–113. Springer, Heidelberg (2013)

12. Law, E., West, K., Mandel, M., Bay, M., Downie, J.S.: Evaluation of algorithms using games: the case of music annotation. In: Proceedings of the 10th International Conference on Music Information Retrieval (ISMIR), pp. 387–392 (October 2009)

13. Jehan, T.: Creating Music by Listening. Ph.D. thesis, Massachusetts Institute of Technology, MA, USA (2005)

14. Wolff, D., Tillman, W.: Adapting metrics for music similarity using comparative judgements. In: Proc. International Symposium on Music Information Retrieval (2011) (accepted for publication)

15. Tsochantaridis, I., Joachims, T., Hofmann, T., Altun, Y.: Large margin methods for structured and interdependent output variables. Journal of Machine Learning Research (JMLR) 6, 1453–1484 (2005)

16. McFee, B., Barrington, L., Lanckriet, G.: Learning similarity from collaborative filters. In: Proceedings of the International Society of Music Information Retrieval Conference, pp. 345–350 (2010)

17. McFee, B., Lanckriet, G.: Heterogeneous embedding for subjective artist similarity. In: Proc. International Symposium on Music Information Retrieval, pp. 513–518 (October 2009)

An Approach to Automatic Music Band Member Detection Based on Supervised Learning

Peter Knees

Department of Computational Perception
Johannes Kepler University, Linz, Austria
peter.knees@jku.at

Abstract. Automatically extracting factual information about musical entities, such as detecting the members of a band, helps building advanced browsing interfaces and recommendation systems. In this paper, a supervised approach to learning to identify and to extract the members of a music band from related Web documents is proposed. While existing methods utilize manually optimized rules for this purpose, the presented technique learns from automatically labelled examples, making therefore also manual annotation obsolete. The presented approach is compared against existing rule-based methods for band-member extraction by performing systematic evaluation on two different test sets.

1 Introduction

Techniques to calculate music similarity are essential for music retrieval and recommendation. In the last years, different content-based methods that capture certain characteristics of a music signal and that are capable of identifying similar pieces with regard to these characteristics have been proposed (for an overview see, e.g., [5]). However, perception of music is a multi-dimensional process that is not only determined by sound properties, but also influenced by cultural and social factors, which cannot be acquired through analysis of the music signal, e.g., advertisements, peer groups, or the media in general. Hence, also similarity between two musical entities can depend on a multitude of factors. A promising approach to deal with the limitations of signal-based methods is to exploit *contextual* information (for an overview see, e.g., [15]). In the majority of existing work, Web- and user-data is used for description/tagging of music (e.g., [10,22,23]) and assessment of similarity between artists (e.g., [16,20,21,25]). However, while for these tasks standard information retrieval (IR) methods that reduce the obtained information to simple representations such as the bag-of-words model may suffice, important information about entities such as artists' full names, band line-up, album and track titles, related artists, as well as some music specific concepts like instrument names and musical styles may be dismissed. By addressing this issue, i.e., by developing methods to identify and extract relevant entities and, in particular, relations between these, essential progress towards improved and multi-faceted similarity measures could be made.

M. Detyniecki et al. (Eds.): AMR 2011, LNCS 7836, pp. 125–139, 2013.

As a consequence, also applications that incorporate similarity measures, such as browsing interfaces and recommendation systems, would benefit and advance.

In this paper, an automatic method to discover a specific semantic relation between musical entities is proposed, namely the *automatic detection of the members of a music band*. More precisely, the task is to determine which persons a music band consists (or consisted) of by analyzing texts from the Web. In the presented first step towards automatic band member detection, no distinction of current or former band members is made, i.e., any person that has been a member of a band at any point in time is considered a band member. In contrast to prior work that addresses the task of extracting band members by utilizing manually determined rule-patterns (see Section 2), here, automatic learning of patterns from labelled data (supervised learning) is proposed. For this, pre-labelled data is required, which is generally difficult to obtain for most types of semantic relations (or rather has to be created still). However, band-membership information is largely available in a structured format (e.g., in the MusicBrainz database[1]) and can therefore be exploited to learn patterns to identify band members also in new items. Furthermore, in contrast to existing rule-based methods, the given approach can be more easily adapted to languages other than English. Currently, however, the focus is on extraction of information from texts written in English.

In the bigger picture, this is supposed to be but the first step towards a collection of methods to identify high-level musical relations. For instance, also Web-based methods to determine relations between music pieces, like cover versions, variations, remasterings, live interpretations, medleys, remixes, samples etc. are conceivable. As some of these concepts are also (partly) deducible from the audio signal itself, ultimately this should result in methods that combine information from the audio with (Web-based) meta-information to automatically discover such relations.

The remainder of this paper is organized as follows. In Section 2, related work from the fields of music information retrieval (MIR) and information extraction (IE) is reviewed. Section 3 explains the details of the proposed approach for band-member extraction as well as the underlying concepts. Section 4 reports on the evaluations carried out. Finally, conclusions are drawn and an outlook over future work is given in Section 5.

2   Related Work

Despite the numerous contributions that exploit Web-based sources to describe music or to derive similarity (cf. Section 1), the number of publications aiming at extracting music-specific factual meta-data is rather small. Schedl et al. [17] propose different approaches to determine the country of origin for a given artist. In one of these approaches, keyword spotting for terms such as "born" or "founded" is performed in the context of countries' names on Web pages. Geleijnse and Korst [7] use patterns like *G bands such as A, for example A_1 and A_2, or M mood*

---

[1] http://musicbrainz.org/

by A (where *G* represents a genre, *A* an artist name, and *M* a possible mood) to unveil genre-artist, artist-artist, and mood-artist relations, respectively.

Band member detection is a specific case of named entity recognition which is itself a well-researched area (for an overview see, for instance, [4]). Named entity recognition comprises the identification of proper names as well as the classification of these names. Existing methods can be classified as either rule-based approaches or supervised learning approaches. While the first type of methods relies on experts that define linguistic rules for the specific task (and domain), the second type requires large amounts of (manually) labelled training data. Alternatively, automatic information extraction can also be driven by an ontology (cf. [2]). Agichtein presents a system that learns to extract relations from unstructured text based on few given examples [1]. In [24], Whitelaw et al. propose a system that automatically generates training data and that is capable of learning fine-grained categories of entities on Web scale document collections. Kim et al. [9] classify named entities using a training corpus automatically labeled by means of a small dictionary and an ensemble of three different learning methods. The supervised learning approach presented in this paper also benefits from knowledge accessible in a structured format that compensates for the requirement of manually labelled data.

With respect to the specific task of band-membershp detection, two rule-based approaches have been presented. In the following, both are reviewed in more detail, as they also serve as references for evaluating the approach presented in this paper.

2.1 Hearst Pattern Approach by Schedl and Widmer

In [18], Schedl and Widmer propose a method to automatically extract the line-up of a music band. In contrast to the work presented in this paper, line-up information includes not only the members of a band but also their corresponding roles, e.g., which instrument they are playing. To obtain documents dealing with a band *B*, Google is invoked with queries such as *B music*, *B music members*, or *B lineup music*. From the retrieved (up to 100 top-ranked) Web pages, n-grams (where $n = \{2, 3, 4\}$), whose tokens consist of capitalized words of length greater than one that are no common speech words, are extracted. For instrument/role detection, a Hearst pattern approach is chosen (cf. [8]). To this end, the following rules are applied to the extracted n-grams and their surrounding text (where *M* is the n-gram/potential band member, *I* an instrument, and *R* a role):

1. *M* plays the *I*
2. *M* who plays the *I*
3. *R M*
4. *M* is the *R*
5. *M*, the *R*
6. *M* (*I*)
7. *M* (*R*)

For I and R, only the roles in the standard rock band line-up, i.e., singer, guitarist, bassist, drummer, and keyboardist, as well as synonyms of these, are taken into consideration. For the final predictions of member-role relations, it is counted on how many Web pages each of the above rules applies (document frequency) and entities that occur on a percentage of the total Web pages that is below a given threshold are discarded.

2.2   Advanced Rule-Patterns by Krenmair

Another rule-based band member extraction approach is presented by Krenmair in [11]. In this work, the GATE framework (General Architecture for Text Engineering; see [6]) is incorporated and more complex rules are defined to automatically identify artist names, to extract band-membership relations, and to extract released albums/media for artists. GATE in an open-source framework that unifies a variety of state-of-the-art text processing and engineering components. With its process-oriented and open architecture, GATE is well-suited for tailoring the included information extraction methods to a specific domain and task. An overview over the general processing pipeline in GATE is given in Section 3.1. To adapt the processing pipeline for the given tasks, Krenmair basically extends two components: the *gazetteer lists* and the so-called *JAPE grammars*[2] used for named entity detection. Gazetteer lists are pre-defined dictionaries of domain-specific entities. For the purpose of detecting musical entities, extensive lists of roles in a band, musical instruments, and musical genres are included. Furthermore, also lists of dates and countries are supplied. The second (and more labor-intense) extension is the generation of grammars for entity detection. For the purpose of band member extraction, a set of rules that consider orthographic features, punctuation, surrounding entities (such as those identified via the gazetteer lists), and surrounding keywords has been designed. For instance, a JAPE grammar rule that aims at finding band members by searching for information about members leaving the band and others joining is given as

```
Rule :  leftJoinedBand (
( ( MemberName ) ) : BandMember
({Token.string == had} | {Token.string == has})?
({Token.string == left} | {Token.string == joined} |
 {Token.string == rejoined} | {Token.string == replaced})
)--> :BandMember.Member = {kind = BandMember, rule = leftJoinedBand}
```

The complete set of JAPE grammars for music-specific entity recognition can be found in Appendix B of [11].

3   Methodology

This section describes the proposed approach of supervised learning for band member detection. Since it makes use of many of the features implemented in

---

[2] JAPE in an acronym for Java Annotation Patterns Engine.

the GATE framework [6], first, an introduction to document processing and named entity detection in GATE is given. Second, the steps undertaken to train a classifier to automatically detect potential band member entities in a text are explained. Finally, a consolidation and filtering step is performed to predict band members.

3.1   Named Entity Recognition in GATE

GATE uses a pipeline architecture for document processing where the sequence and composition of processing resources (PR) can be adapted in order to suit the specific task. In each processing step, the corresponding PR creates or modifies annotations in the text that are passed to the subsequent PRs. Typically, a GATE pipeline consists of the following PRs:

1. **Tokenizer**: splits the text into tokens based on white spaces.
2. **Sentence Splitter**: splits the text into sentences based on punctuation.
3. **Part of Speech Tagger**: assigns part-of-speech (PoS) tags to tokens, i.e., annotates each token with it's linguistic category (noun, verb, preposition, etc.).
4. **Gazetteer**: a collection of word lists/dictionaries that are compiled into finite state machines. This is used for named entity look-up (see below).
5. **Transducer**: aims at identifying named entities using manually generated JAPE grammar rules. These rules can include lexical expressions, PoS information, entities extracted via the gazetteer, or any other type of available annotation (cf. Section 2.2).
6. **Orthografic Matching**: finds identities among named entities

GATE includes PRs for all of these steps and provides therefore rule-based named entity recognition and detection of persons in texts out-of-the-box. It should be noted that the process of person detection is interwoven with the detection of other entities. Nevertheless, the following outlines the particularities of the person detection process: Using a gazetteer, first names and titles are identified. In the transducer step, initials, first names, surnames, and endings are detected, for instance, by using orthographic characteristics (e.g., capitalization) and PoS information. This information is then combined with the information obtained from the gazetteers. In a post-processing step, persons' surnames are removed if they contain certain stopwords or can be attributed to an organization. Details about this can be found in Appendix F of the GATE User Guide[3].

3.2   Extracting Band Members by Supervised Learning

Construction of rules for the transducer step is a tedious work. The more heterogeneous the underlying data is, the more special cases have to be covered. The idea to alleviate this is to apply a supervised learning algorithm to a set of user-annotated examples. Using the learned model, relevant information could be

---

[3] http://gate.ac.uk/userguide/

extracted also from new documents. Several approaches, more precisely, several types of machine learning algorithms, have been proposed for information extraction tasks, such as hidden-markov-models [3], decision trees [19], or support vector machines (SVM) [12]. The GATE framework offers a machine learning PR that supports various types of classification algorithms [13]. Since examples in the literature (e.g., [12]) show that SVMs may yield results that rival those of rule-based approaches, SVMs are chosen as classifier.

Training Data. For training of the SVMs, a set of annotated documents is required. To this end, the set of 83 artists/bands that was used as training set in [11] is utilized. Since there is a small overlap of bands from this set with one of the evaluations sets (i.e., the Metal page set), bands that also occur in the evaluation set were removed from the training set. For the remaining bands, informative texts, i.e., band biographies, are obtained via the echonest API[4]. Using the echonest's Web service, related biographies (e.g., from Wikipedia[5], last.fm[6], allmusic[7], or Aol Music[8]) can be conveniently retrieved in plain text format. Since among the provided biographies for a band, duplicates or near-duplicates, as well as only short snippets can be observed, (near-)duplicates as well as biographies consisting of less than 100 characters are filtered out. Furthermore, all biographies consisting of over 40 kilobytes of data are removed to keep processing times short. In total, a set of 126 documents remains. The corresponding ground truth for these bands, i.e., the actual list of current and former band members, is derived by consulting MusicBrainz and the bands' Wikipedia pages.

To annotate the 126 documents to serve as training examples for the SVM, labeling is performed in two steps. First, documents are annotated using the standard GATE pipeline (see Section 3.1) extended by the gazetteer lists used by Krenmair (see Section 2.2). Thus, also potential person annotations are obtained using the named entity functionality. In the second step, the detected persons are compared against the elements of the band's ground truth and annotated as band member if they match one of the elements or one of the elements' last token (to annotate band members that are only referred to by their last names).

Feature Construction. Construction of the features for SVM training is carried out as described by Li et al. [12]. Following their approach two distinct SVM classifiers are trained to detect Person entities to be marked as band members. The first classifier aims at predicting the beginning of a band member entity (i.e., to classify whether a token is the first token of a band member's name), whereas the second aims at predicting the end (i.e., whether a token is the last token of a band member's name). From the obtained predictions of start and end positions,

---

[4] http://developer.echonest.com
[5] http://www.wikipedia.org
[6] http://www.last.fm
[7] http://www.allmusic.com
[8] http://music.aol.com

actual members, as well as corresponding confidence scores are determined in a post-processing step. In the following, a comprehensive description is given (for more details the reader is referred to the original sources [12, 13]).

Prior to classifying a token, a feature vector representation has to be obtained. In the given scenario, for each token, its content (i.e., the actual string), orthographic properties (such as capitalization), PoS information (conjunctions, verbs, nouns, determiners, etc. – in total over 50 different tags), and gazetteer-based entity information (e.g., dates, locations, genre) are considered. Person annotations that are marked as band members serve as target class. To gather all feature attributes, the training corpus is scanned for all occurring values of any of these annotations. Then, for each token a feature vector is constructed where each potential value corresponds to one dimension which is set to 1 if the token is annotated with the corresponding value. In addition also the context of each token (consisting of a window that includes the 5 preceding and the 5 subsequent tokens) is incorporated. This is achieved by creating an SVM input vector for each token that is a concatenation of the feature vectors of all tokens in the context window. To reflect the distance of the surrounding tokens to the actual token (i.e., the center of the window), a reciprocal weighting is applied, meaning that "the nonzero components of the feature vector corresponding to the jth right or left neighboring word are set to be equal to $1/j$ in the combined input vector." [12]. In our experiments, this results in feature vectors with about 1.5 million dimensions.

For SVM training, every single token of all text documents in the training corpus (its input vector, rather) serves as example — once for learning to identify start tokens of persons that are band members and once for learning to identify end tokens. To deal with the unbalanced distribution of positive and negative training examples, a special form of SVMs is used, namely an SVM with uneven margins [14].

Entity Extraction. After classifying individual tokens into start and/or end tokens, a post-processing technique is applied to detect band members and assign a confidence score. First, start tokens without matching end token, as well as end tokens without matching start token are removed. Second, entities with a length (in terms of the number of tokens) that does not match any training example's length are discarded. Third, a confidence score is calculated based on a probabilistic interpretation of the SVM output for all possible classes. More precisely, for each entity, the Sigmoid transformed SVM output probabilities of start and end token are multiplied for each possible output class. Finally, the class (label) with the highest probability is predicted for the entity if its probability is greater than 0.25. The probability of the predicted class serves also as a confidence score.

As a result, an information extraction resource is obtained that processes texts and outputs potential band member entities as well as corresponding confidence scores. To evaluate the impact of the number of training examples, two SVM classifiers are trained – one using all 126 documents and one using a random subset of 50 documents.

3.3 Entity Consolidation and Member Prediction

From the named entity extraction step, for each processed text, a list of potential band members is obtained. For each band, the lists from all texts associated with the band are joined and the occurrences of each entity as well as the number of texts an entity occurs in are counted. The resulting collection contains a lot of noise, making a filtering and merging step necessary. First, all entities with a confidence score below 0.5 are removed since they are more likely to not represent band members than representing band members according to the classification step. On the cleaned list, the same observations as described in [18] can be made, namely that some members are referenced with different spellings (*Paavo Lötjönen* vs. *Paavo Lotjonen*), with abbreviated first names (*Phil Anselmo* vs. *Philip Anselmo*), with nicknames (*Darrell Lance Abbott* vs. *Dimebag Darrell* or just *Dimebag*), or only by their last name (*Iommi*). As in [18], this is dealt with by introducing an approximate string matching function, namely the level-two Jaro-Winkler similarity.[9] According to [18], this type of similarity function is suited for comparing names as it assigns higher matching scores to pairs of strings that start with the same sequence of characters. In the level-two variant, the two entities to compare are split into substrings and similarity is calculated as an aggregated similarity of pairwise comparison of the substrings. On the list of extracted band members, two entities are considered synonymous if their level-two Jaro-Winkler similarity is above 0.9. In addition, to deal with the occurrence of last names, an entity consisting of one token is considered a synonym of another entity if it matches the other entity's last token.

This consolidated list is usually still noisy, calling for additional filtering steps. To this end, two threshold parameters are introduced. Using the first threshold, $t_f \in \mathbb{N}^0$, the minimum number of occurrences of an entity (or its synonyms) to be predicted is determined. The second threshold, $t_{df} \in [0...1]$ controls the lower bound of the fraction of texts/documents associated with the band an entity has to occur in (document frequency in relation to the total number of documents per band). The impact of these two parameters is systematically evaluated in the following section.

4 Evaluation

To assess the potential of the proposed approach, to compare it with existing approaches, and to measure the impact of the parameters, systematic experiments are conducted. This section details the used test collections as well as the applied evaluation measures and reports on the results of the experiments.

4.1 Test Collections

For evaluation, two collections with different characteristics are used. The first collection is a set of Web pages introduced in [18]. This set consist of Google's 100

---

[9] For similarity calculation, the open-source Java toolkit *SecondString* (`http://secondstring.sourceforge.net`) is utilized.

top-ranked Web pages retrieved using the query *"band name" music members* (cf. Section 2.1) for 51 Rock and Metal bands (resulting in a total of 5,028 Web pages). In [18], this query setting yielded best results and is therefore chosen as reference. As a ground truth, the membership-relations that include former members are chosen (i.e., the M_f ground truth set of [18]). For this evaluation collection also the results obtained by applying the Hearst patterns proposed by Schedl and Widmer are available, allowing for a direct comparison of the approaches' band member extraction capabilities.

The second test collection is a larger scale collection consisting only of band biographies to be found on the Web. Starting from a snapshot of the MusicBrainz database from December 2010, all artists marked as bands and all corresponding band members are extracted.[10] In addition, for these bands, also band-membership information from Freebase[11] is retrieved and merged with the MusicBrainz information to make the ground truth data set more comprehensive. After this step, band-membership information is available for 34,238 bands. As with the training set, for each band name, the echonest API is invoked to obtain related biographies. After filtering (near-)duplicates and snippets, for 23,386 bands (68%) at least one biography remains. In total, a set of 38,753 biographies is obtained. In comparison to the first test collection, i.e., Schedl's Metal page set, the biography set contains more bands, more specific documents in a homogeneous format (i.e., biographies instead of semi-structured Web pages from various sources), but less associated documents (in average 1.66 documents per band, as opposed to an average of 98.5 documents per band for the Metal page set).

4.2   Evaluation Metrics

For evaluation, *precision*, *recall*, and *F-measure* (i.e., the harmonic mean of precision and recall) are calculated separately for each band and averaged over all bands to obtain a final score. The metrics are defined as follows:

$$precision = \begin{cases} \frac{|T \cap P|}{|P|} & \text{if } |P| > 0 \\ 1 & \text{otherwise} \end{cases} \tag{1}$$

$$recall = \frac{|T \cap P|}{|T|} \tag{2}$$

$$F = 2 \cdot \frac{precision \cdot recall}{precision + recall} \tag{3}$$

where P is the set of predicted band members and T the ground truth set of the band. To assess whether an extracted band member candidate is correct, again the level-two Jaro-Winkler similarity (see Section 3.3) is applied. More precisely, if the Jaro-Winkler similarity between a predicted band member and a member

---

[10] Bands contained in the training set are excluded.
[11] http://www.freebase.com

contained in the ground truth is greater than 0.9, the prediction is considered to be correct. Furthermore, if a predicted band member name consist of only one token, it is considered correct, if it matches with the last token of a member in the ground truth. This weakened definition of matching allows for tolerating small spelling variations, name abbreviations, extracted last names, as well as string encoding differences (cf. [18]).

For comparison with Schedl's Hearst patterns on the Metal page set, it has to be noted that in [18], calculation of precision and recall is done on the full set of bands and members (and their corresponding roles), yielding global precision and recall values, whereas here, the evaluation metrics are calculated separately for each band and are then averaged over all bands to remove the influence of a band's size. Using the global evaluation scheme, e.g., orchestras are given far more importance than, for instance, duos in the overall evaluation, although for a duo, the individual members are generally more important than for an orchestra. Therefore, in the following, the different approaches are compared based on macro-averaged evaluation metrics (calculated using the arithmetic mean of the individual results).

4.3   Evaluation Results

To gain insights into the applicability of the proposed supervised learning approach (denoted as SVM), it is compared with a baseline consisting of the out-of-the-box person identification function implemented in GATE (Section 3.1), with the advanced rule-pattern approach by Krenmair (Section 2.2), and — on the Metal page set — also with Schedl's Hearst pattern approach (Section 2.1). In addition, also the upper bound for the recall is calculated. This upper bound is implied by the underlying documents, since band members that do not occur on any of the documents can not be predicted (cf. [18]).

Figure 1 shows Precision-Recall curves for the different band member detection approaches on the Metal page set. For a systematic comparison with Schedl's Hearst pattern approach, the t_{df}, i.e., the threshold that determines on which fraction of a band's total documents a band member has to appear on to be predicted, is varied. It can be seen that the advanced rule-based approach clearly performs best. Also the supervised learning approaches (SVM with 126 and 50 pages to learn from) outperform the Hearst pattern approach. It becomes apparent that on the Metal set, advanced rule patterns, the GATE person detection, and the supervised approaches can yield recall values close to the upper bound, i.e., these approaches capture nearly all members contained in the documents at least once. For the Hearst patterns, recall remains low. The impression that GATE person detection and Hearst patterns perform worse on the Metal page set than the SVM approaches and that the manually tailored rules yield by far the best results is further supported by the maximum F-measure values given in Table 1. However, when comparing the Hearst patterns by Schedl and Widmer, it has to be noted that their approach was initially designed to also detect the roles of the band members — a feature that none of the other evaluated approaches is capable of.

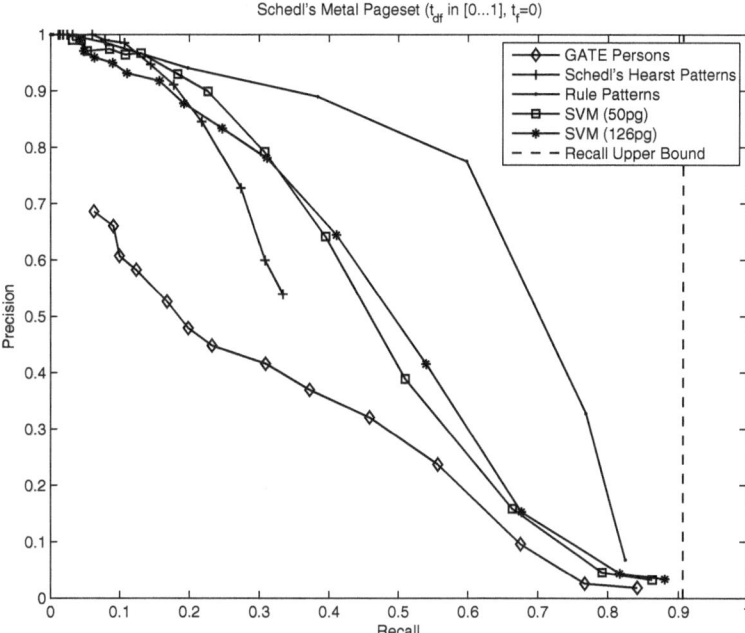

Fig. 1. Precision-Recall plots for comparing the learning-based approach with the rule-based approaches on the Metal page set from [18]. Curves are obtained by systematically varying the t_{df} parameter in the range of 0 to 1 in steps of 0.1 and averaging precision and recall over all 51 bands.

Table 1. Maximum F-measure values and corresponding settings on the Metal page set from [18]. Values are obtained by averaging over all 51 bands.

	Settings	**F-Measure**
GATE Persons	$t_{df} = 0.8$, $t_f = 2$	0.39
Hearst Patterns	$t_{df} = 0.0$	0.41
Rule Patterns	$t_{df} = 0.05$, $t_f = 0$	0.67
SVM (50 pages)	$t_{df} = 0.15$, $t_f = 13$	0.49
SVM (126 pages)	$t_{df} = 0.15$, $t_f = 1$	0.50

Since on the biography set only 1.66 documents per band are available on average, variation of the t_{df} threshold is not as interesting as on the Metal page set. Therefore, Figure 2 depicts curves of the approaches with varying values of t_f, i.e., the threshold that determines how often an entity has to be detected to be predicted as a band member. On this set, the supervised learning approaches outperform the rule-based extraction approach. In contrast to the Metal page set, there seems to be no difference between the SVMs trained on 50 and 126 documents, respectively. Also, it can be seen that the supervised

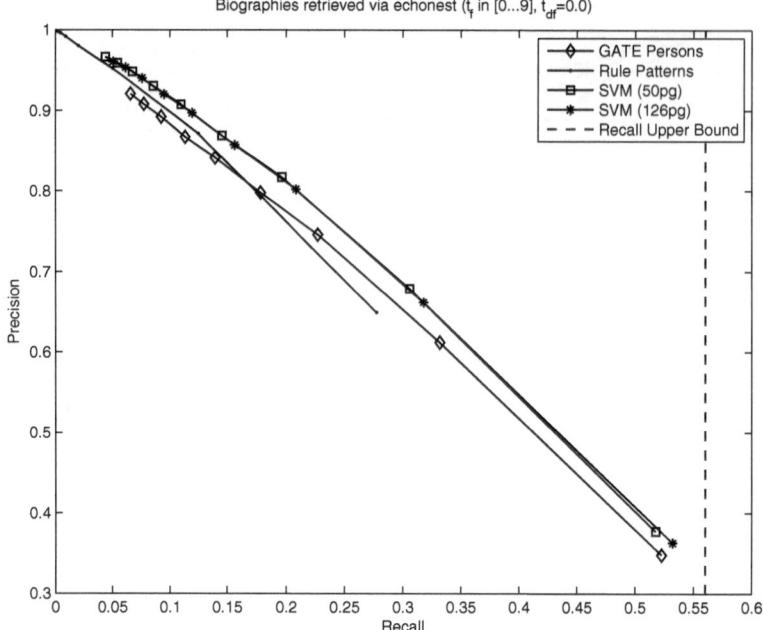

Fig. 2. Precision-Recall plots for comparing the learning-based approach with the advanced rule-based approach on the biography set. Curves are obtained by systematically varying the t_f parameter in the range of 0 to 9 in steps of 1 and averaging precision and recall over all 23,386 bands.

Table 2. Maximum F-measure values and corresponding settings on the biography set. Values are obtained by averaging over all 23,386 bands.

	Settings	F-Measure
GATE Persons	$t_{df} = 0.8$, $t_f = 0$	0.44
Rule Patterns	$t_{df} = 0.0$, $t_f = 0$	0.39
SVM (50 pages)	$t_{df} = 0.65$, $t_f = 0$	0.45
SVM (126 pages)	$t_{df} = 0.6$, $t_f = 0$	0.45

learning approaches exhibit a behavior similar to the GATE person detection baseline with only slightly better precision values. Also from the maximum F-measure achieved by these approaches, it can be seen that there is only a marginal difference (cf. Table 2). A finding that is consistent for both collections is that F-measure values of around 0.5 can be expected using the SVM approaches. Also on both collections it can be observed that the GATE person detection yields best results with high values of t_{df}. i.e., when relying on a larger amount of evidence.

4.4 Discussion of Results

The observations that can be made are not consistent on the two collections. On the Metal set, the advanced rule-based approach outperforms the supervised learning approaches clearly. On the biography set, supervised learning approaches perform better. The obvious explanation for this behavior is that the SVMs have been trained on biographies, whereas the rule-patterns have been generated based on human observations. Without doubt, SVMs (and all other supervised learning approaches) benefit from similarly structured input in both training and test set. In this case, also a classifier trained on a smaller set of documents can yield nearly identical results. Moreover, biographies typically follow a certain writing style and consist — in contrast to arbitrary Web pages — mostly of grammatically well-formed sentences. Clearly, natural language processing techniques such as PoS tagging perform best on this type of input. This seems also to be the reason why the standard GATE person detection approach works well on the biography data, but inferiorly on the Metal page set.

5 Conclusions and Future Work

In this paper, an approach to band member extraction from Web documents that uses supervised learning has been proposed. While it became evident that on heterogeneous data sources manually generated rules are yielding superior results in terms of precision, it could be seen that supervised approaches are a particularly good choice when dealing with many documents of similar structure.

In general, the results obtained show great potential for this and also related tasks. For instance, just by focusing on biographies, a lot of highly relevant meta-information on music could be extracted. For instance, consider the following paragraph taken from the Wikipedia page of *Brendan Benson*:

"Also in 2003, Benson released an EP, Metarie, with his then band The Wellfed Boys. The EP featured a cover of Paul McCartney's "Let Me Roll It" which featured back-up vocals by friend and later fellow member of The Raconteurs; Jack White."[12]

This short paragraph contains discography information for *Brendan Benson*, information on membership in two bands (*The Wellfed Boys* and *The Raconteurs*) and further line-up information for *The Raconteurs*. This allows to infer relations between the mentioned bands, as well as the mentioned persons. In addition, this paragraph informs that *Paul McCartney* is the composer of the song *Let Me Roll It*, that *Brendan Benson* has covered this song, and that *Jack White* appeared as vocalist on the recording. Using further information extraction methods, in future work, it should be possible to capture at least some of this semantic information and relations and to advance the current state-of-the-art in music retrieval and recommendation.

---

[12] http://en.wikipedia.org/w/index.php?title=Brendan_Benson&oldid=447778757

Acknowledgments. Thanks are due to Andreas Krenmair for sharing his implementation of the JAPE grammars and to Markus Schedl for making available his data set and also for his support with the evaluations. Johann Petrak deserves credit for his support with the GATE framework. This research is supported by the Austrian Research Fund (FWF) under grant L511-N15.

References

1. Agichtein, Y.: Extracting relations from large text collections. Ph.D. thesis, Columbia University, New York, NY, USA (2005)
2. Alani, H., Kim, S., Millard, D.E., Weal, M.J., Hall, W., Lewis, P.H., Shadbolt, N.R.: Automatic Ontology-Based Knowledge Extraction from Web Documents. IEEE Intelligent Systems 18(1), 14–21 (2003)
3. Bikel, D.M., Miller, S., Schwartz, R., Weischedel, R.: Nymble: a High-Performance Learning Name-finder. In: Proceedings of the 5th Conference on Applied Natural Language Processing, pp. 194–201 (1997)
4. Callan, J., Mitamura, T.: Knowledge-Based Extraction of Named Entities. In: Proceedings of the 11th International Conference on Information and Knowledge Management (CIKM 2002), pp. 532–537. ACM, New York (2002)
5. Casey, M.A., Veltkamp, R., Goto, M., Leman, M., Rhodes, C., Slaney, M.: Content-Based Music Information Retrieval: Current Directions and Future Challenges. Proceedings of the IEEE 96, 668–696 (2008)
6. Cunningham, H., Maynard, D., Bontcheva, K., Tablan, V.: GATE: A framework and graphical development environment for robust NLP tools and applications. In: Proceedings of the 40th Anniversary Meeting of the Association for Computational Linguistics (ACL 2002), Philadelphia (July 2002)
7. Geleijnse, G., Korst, J.: Web-based artist categorization. In: Proceedings of the 7th International Conference on Music Information Retrieval (ISMIR 2006), Victoria, Canada (October 2006)
8. Hearst, M.A.: Automatic acquisition of hyponyms from large text corpora. In: Proceedings of the 14th Conference on Computational Linguistics (COLING 1992), vol. 2, pp. 539–545. Association for Computational Linguistics, Stroudsburg (1992)
9. Kim, J.H., Kang, I.H., Choi, K.S.: Unsupervised named entity classification models and their ensembles. In: Proceedings of the 19th International Conference on Computational Linguistics (COLING 2002), vol. 1, pp. 1–7 (2002)
10. Knees, P.: Text-Based Description of Music for Indexing, Retrieval, and Browsing. Ph.D. thesis, Johannes Kepler Universität, Linz, Austria (November 2010)
11. Krenmair, A.: Musikspezifische Informationsextraktion aus Webdokumenten. Diplomarbeit, Johannes Kepler University, Linz, Austria (May 2010)
12. Li, Y., Bontcheva, K., Cunningham, H.: SVM Based Learning System for Information Extraction. In: Winkler, J.R., Niranjan, M., Lawrence, N.D. (eds.) Deterministic and Statistical Methods in Machine Learning. LNCS (LNAI), vol. 3635, pp. 319–339. Springer, Heidelberg (2005)
13. Li, Y., Bontcheva, K., Cunningham, H.: Adapting SVM for Data Sparseness and Imbalance: A Case Study on Information Extraction. Natural Language Engineering 15(2), 241–271 (2009)
14. Li, Y., Shawe-Taylor, J.: The SVM with uneven margins and Chinese document categorization. In: Proceedings of The 17th Pacific Asia Conference on Language, Information and Computation (PACLIC 17), pp. 216–227 (2003)

15. Schedl, M., Knees, P.: Context-based Music Similarity Estimation. In: Proceedings of the 3rd International Workshop on Learning the Semantics of Audio Signals (LSAS 2009), Graz, Austria (December 2009)

16. Schedl, M., Knees, P., Widmer, G.: A Web-Based Approach to Assessing Artist Similarity using Co-Occurrences. In: Proceedings of the 4th International Workshop on Content-Based Multimedia Indexing (CBMI 2005), Riga, Latvia (2005)

17. Schedl, M., Schiketanz, C., Seyerlehner, K.: Country of Origin Determination via Web Mining Techniques. In: Proceedings of the IEEE International Conference on Multimedia and Expo (ICME 2010): 2nd International Workshop on Advances in Music Information Research (AdMIRe 2010), Singapore, July 19-23 (2010)

18. Schedl, M., Widmer, G.: Automatically Detecting Members and Instrumentation of Music Bands via Web Content Mining. In: Boujemaa, N., Detyniecki, M., Nürnberger, A. (eds.) AMR 2007. LNCS, vol. 4918, pp. 122–133. Springer, Heidelberg (2008)

19. Sekine, S.: NYU: Description of the Japanese NE system used for MET-2. In: Proceedings of the 7th Message Understanding Conference (MUC-7) (1998)

20. Shavitt, Y., Weinsberg, U.: Songs Clustering Using Peer-to-Peer Co-occurrences. In: Proceedings of the IEEE International Symposium on Multimedia (ISM 2009): International Workshop on Advances in Music Information Research (AdMIRe 2009), San Diego, CA, USA (December 2009)

21. Slaney, M., White, W.: Similarity Based on Rating Data. In: Proceedings of the 8th International Conference on Music Information Retrieval (ISMIR 2007), Vienna, Austria (September 2007)

22. Sordo, M., Laurier, C., Celma, O.: Annotating Music Collections: How Content-based Similarity Helps to Propagate Labels. In: Proceedings of the 8th International Conference on Music Information Retrieval (ISMIR 2007), Vienna, Austria, pp. 531–534 (September 2007)

23. Turnbull, D., Barrington, L., Lanckriet, G.: Five Approaches to Collecting Tags for Music. In: Proceedings of the 9th International Conference on Music Information Retrieval (ISMIR 2008), Philadelphia, PA, USA (2008)

24. Whitelaw, C., Kehlenbeck, A., Petrovic, N., Ungar, L.: Web-scale named entity recognition. In: Proceedings of the 17th ACM Conference on Information and Knowledge Management (CIKM 2008), pp. 123–132 (2008)

25. Whitman, B., Lawrence, S.: Inferring Descriptions and Similarity for Music from Community Metadata. In: Proceedings of the 2002 International Computer Music Conference (ICMC 2002), Gothenburg, Sweden, pp. 591–598 (September 2002)

Author Index